GOD'S DESIGN®

Heaven & Earth
for beginners

| Weather and Water for Beginners | Universe for Beginners | Planet Earth for Beginners |

MASTERBOOKS® CURRICULUM

Debbie & Richard Lawrence

MasterBooks® CURRICULUM

Authors: Debbie and Richard Lawrence

Master Books Creative Team:
Editor: Laura Welch
Design: Jennifer Bauer
Cover Design: Diana Bogardus

Copy Editors:
Judy Lewis
Willow Meek

Curriculum Review:
Kristen Pratt
Laura Welch
Diana Bogardus

First printing: August 2020
Third printing: July 2022

Master Books, P.O. Box 726, Green Forest, AR 72638
Master Books® is a division of the New Leaf Publishing Group, Inc.

ISBN: 978-1-68344-238-7
ISBN: 978-1-61458-760-6 (Digital)

Printed in the United States of America

Please visit our website for other great titles:
www.masterbooks.com

NOTE: All activities and projects should be supervised by an adult. The publisher and authors have made every reasonable effort to ensure that the activities recommended in this book are safe when performed as instructed but assume no responsibility for any damage caused or sustained while conducting the experiments and activities. It is the parents', guardians', and/or teachers' responsibility to supervise all recommended activities.

About the Authors:

The God's Design Science series is based on a biblical worldview and reveals how science supports the biblical account of creation. **Richard and Debbie Lawrence**, authors of the series, have a long history of enjoying science. They have both worked as electrical engineers and now Debbie teaches chemistry and physics at a homeschool co-op. While homeschooling their children for 16 years, there was almost always a science experiment going on in the kitchen. Today that tradition is being continued with the next generation as the grandkids enjoy Grandma's Science Day once a week.

Weather and Water for Beginners

Universe for Beginners

Planet Earth for Beginners

Teacher Introduction

Welcome to God's Design Science for Beginners! *God's Design for Heaven and Earth* includes Weather and Water, Universe, and Planet Earth.

The textbook is organized into three units of 35 lessons each. These lessons include a section to read, vocabulary words to trace, review questions, and one or more activities. These may be a pen-on-paper activity, a Scripture verse to trace, or a hands-on experiment. Each unit concludes with a vocabulary review.

In the back of this book, you will find answer keys for puzzles, review questions, and additional activities and experiments you can do with your student. We encourage you to do as many additional activities and experiments as possible since children learn more, and retain more, when they actively engage with the material.

The supply list for each section is included on pages 8–10. Be sure to consult the lists and gather the supplies you'll need in advance.

There is no set timeline for completing each book. You can decide how many days each week you want to do science. We suggest that you review previous lessons often to help your students retain the information they've learned.

This course also includes unit vocabulary reviews. These can be used as assessments when given as a quiz or test if you wish.

A course schedule is included in this book. And as always, you can adjust it per the needs of your student. All activities can also be modified as needed to best fit your school year per your discretion.

We hope that you have a wonderful time of discovery as you explore the history of the planet earth — exploring rocks, volcanoes, and glaciers, as well as learning about our world's climate, clouds, types of weather, oceans, and also about the wonderful universe God created — planets, meteors, solar energy, and more!

If you are using both the God's Design Science for Beginners book in conjunction with the *God's Design for Heaven and Earth* course for older students to create a multi-age level study for your students, please keep in mind the following tips:

1	2	3
Be flexible with the schedule for your students. Adjust as needed.	Focus on the ability of each student to master the material and make it fun!	Encourage students to help one another as they learn the material.

You can download a schedule that will help you teach this course at the same time as the course for the older student. It is available at www.nlpg.com/classroom-aids.

Watch for the added activities symbol — these give you additional hands-on learning opportunities.

SEE OPTIONAL ACTIVITY

Special Project

At the end of this course, the student will have the opportunity to do a special project. This can be:

✓ A poster sharing something you have learned from the course (e.g., focused on the earth, the universe, or weather)

✓ A short report from any of the three sections — planet earth, weather and water, or the universe. Can be focused on a specific area of interest (e.g., rocks, geologic processes, types of weather, weather events in history, the planets, the solar system, etc).

✓ A short oral presentation for your teacher: explain what you enjoyed most about the course and why.

Be as creative as you want to be!

Weather and Water for Beginners – Supply List

Lesson	✔	Supplies
2		Small candle, glass jar, modeling clay
3		Balloons, string, yardstick, tape
4		Paper plate, weather symbols from the lesson, glue
5		Construction paper, pictures of various climates, yarn
7		An old calendar
8		Yellow finger paint
10		Glass jar with lid, pan, plastic zip bag, ice
11		Blue construction paper, cotton balls, glue
12		White paper
14		Metal hanger, trash bag, tape
15		Stuffed animal, cloth
16		Two 2–liter soda bottles, tornado tube connector (available from many science suppliers) or duct tape (works but may leak)
18		Thermometer
19		Clear jar, masking tape, permanent marker, ruler
22		Glass bottle, balloon, pan
24		Dark–colored construction paper, salt water, paint brush
25		Pie pan, water, pepper, straw
26		Sink or bathtub, small bottle, large bottle
28		Paint roller pan/tray with ramp, empty plastic bottle, sand
29		Several colors of sand or a box of fruit ring cereal, baby food jars
31		Empty aquarium, fish bowl or other glass or clear plastic container, modeling clay or play dough
32		Mural or butcher block paper, crayons, water color paints
33		Gummy worms, ice cream, whipped cream, chocolate syrup
34		Modeling clay or play dough

Lesson	✔	Supplies
3		Access to a recording of Vivaldi's Four Seasons. Available on the Internet.
4		Magnifying glass, telescope (optional)
5		Dark construction paper, white crayon, star stickers
6		Tagboard (heavy–weight cardboard) and string, or glow in the dark stars (optional)
7		Black construction paper, glue, sand (or glitter)
8		A variety of rocks, ball
9		Styrofoam™ ball, tagboard, glue, glitter
12		Pie pan, small mirror
13		Sidewalk chalk
14		Globe or large ball, small ball, flashlight
16		Sand box sand, corn starch
17		Sandwich cookies (like Oreos™)
19		Towel, hair dryer, ice
21		6 sided die
23		Two cereal bowls, marbles or round cereal
25		Apple, popsicle stick
27		Blocks, tape measure
29		Balloon
30		Round crackers, graham crackers, peanut butter
32		Waxed paper, butter knife
33		Winter clothes such as coat, snow pants, boots and gloves, bike helmet
34		Modeling clay – optional – you can use a Styrofoam™ or plastic kit instead of the modeling clay.

Planet Earth for Beginners – Supply List

Lesson	✔	Supplies
4		Large jar, sand, dirt, pebbles, rocks, twigs, dried leaves
5		2 bowls, ice cubes
6		Glass container, ice cubes
7		Bowl, dirt or pebbles, ice cubes
8		Apple
10		Old crayons, muffin pan, foil cup liners
11		Rice Krispies®, candy pieces, raisins, marshmallows, butter, sidewalk chalk
12		Modeling clay or play dough, plaster of Paris, small toy animal or a seashell
14		Old crayons, aluminum foil, oven mitts, hair dryer
16		Many different rocks, rocks and minerals guidebook
19		Globe, baking sheet, frosting, graham crackers
20		Modeling clay or play dough
21		Newspaper or paper towels
22		Building blocks
23		Items for an emergency — suggestions from the lesson include: non-perishable food (such as dried fruit or peanut butter), can opener (manual), first aid kit, flashlight with extra batteries, matches, toothbrush, toothpaste, soap, paper plates, plastic cups and utensils, paper towels, water — at least a gallon per person, per day — sleeping bag or warm blanket for each person
24		Bottle, baking soda, vinegar, cookie sheet or pan, modeling clay or play dough
25		Ice cream, chocolate syrup, cookie crumbs; alternative supplies: mashed potatoes, gravy, bread crumbs
27		Soda straws
28		Baking sheet, dirt or sand
29		Baking sheet, dirt or potting soil
30		Baking sheet
31		Potting soil, yard soil, magnifying glass
33		Salt, dark construction paper
34		A container with many different sections for displaying a rock collection, rocks, rocks and minerals guidebook

Date	Day	Assignment	Due Date	✓	Grade
		First Semester-First Quarter			
Week 1	Day 1	Weather and Water for Beginners Unit 1: Atmosphere and Meteorology Do Lesson 1: God Made Weather • Pages 20–22			
	Day 2	Do Lesson 2: The Atmosphere • Pages 23–25			
	Day 3	Do Lesson 3: The Weight of Air • Pages 26–27			
	Day 4				
	Day 5				
Week 2	Day 6	Do Lesson 4: The Study of Weather • Pages 28–31			
	Day 7	Complete **Atmosphere and Meteorology Unit Vocabulary Review** (Lessons 1–4) • Page 32			
	Day 8	Weather and Water for Beginners Unit 2: Ancient Weather and Climate Do Lesson 5: Weather vs. Climate • Pages 34–36			
	Day 9				
	Day 10				
Week 3	Day 11	Do Lesson 6: Climate Before the Flood • Pages 37–38			
	Day 12	Do Lesson 7: The Great Flood • Pages 39–41			
	Day 13	Do Lesson 8: Climate After the Flood • Pages 42–43			
	Day 14	Complete **Ancient Weather and Climate Unit Vocabulary Review** (Lessons 5–8) • Page 44			
	Day 15				
Week 4	Day 16	Weather and Water for Beginners Unit 3: Clouds Do Lesson 9: Water Cycle • Pages 46–47			
	Day 17	Do Lesson 10: Forming Clouds • Pages 48–50			
	Day 18	Do Lesson 11: Cloud Types • Pages 51–53			
	Day 19	Do Lesson 12: Precipitation • Pages 54–55			
	Day 20				
Week 5	Day 21	Complete **Clouds Unit Vocabulary Review** (Lessons 9–12) Page 56			
	Day 22	Weather and Water for Beginners Unit 4: Storms Do Lesson 13: Air Masses and Weather Fronts • Pages 58–59			
	Day 23	Do Lesson 14: Wind • Pages 60–62			
	Day 24	Do Lesson 15: Thunderstorms • Pages 63–65			
	Day 25				
Week 6	Day 26	Do Lesson 16: Tornadoes • Pages 66–68			
	Day 27	Do Lesson 17: Hurricanes • Pages 69–71			
	Day 28	Complete **Storms Unit Vocabulary Review** (Lessons 13–17) • Page 72			
	Day 29	Weather and Water for Beginners Unit 5: Weather Information Do Lesson 18: Measuring Temperature and Air Pressure • Pages 74–76			
	Day 30				

Date	Day	Assignment	Due Date	✓	Grade
Week 7	Day 31	Do Lesson 19: Measuring Rainfall and Wind Speed • Pages 77–79			
	Day 32	Do Lesson 20: Predicting Weather • Pages 80–81			
	Day 33	Do Lesson 21: Weather Sayings • Pages 82–84			
	Day 34	Do Lesson 22: Weather Review • Pages 85–87			
	Day 35				
Week 8	Day 36	Complete **Weather Information Unit Vocabulary Review** (Lessons 18–22) • Page 88			
	Day 37	Weather and Water for Beginners Unit 6: Ocean Movements Do Lesson 23: Oceans • Pages 90–92			
	Day 38	Do Lesson 24: Why Is Seawater Salty? • Pages 93–94			
	Day 39	Do Lesson 25: Ocean Currents • Pages 95–97			
	Day 40				
Week 9	Day 41	Do Lesson 26: Waves • Pages 98–100			
	Day 42	Do Lesson 27: Tides • Pages 101–103			
	Day 43	Do Lesson 28: Wave Erosion • Pages 104–105			
	Day 44	Do Lesson 29: Building Beaches • Pages 106–108			
	Day 45				
		First Semester-Second Quarter			
Week 1	Day 46	Complete **Ocean Movements Unit Vocabulary Review** (Lessons 23–29) • Pages 109–110			
	Day 47	Weather and Water for Beginners Unit 7: Seafloor Do Lesson 30: Sea Exploration • Pages 112–114			
	Day 48	Do Lesson 31: The Ocean Floor • Pages 115–117			
	Day 49	Do Lesson 32: Ocean Zones • Pages 118–120			
	Day 50				
Week 2	Day 51	Do Lesson 33: Vents and Smokers • Pages 121–123			
	Day 52	Do Lesson 34: Coral Reefs • Page 124–126			
	Day 53	Do Lesson 35: Conclusion • Pages 127–128			
	Day 54	Complete **Seafloor Unit Vocabulary Review** (Lessons 30–35) • Pages 129–130			
	Day 55				
Week 3	Day 56	Universe for Beginners Unit 1: Space Models and Tools Do Lesson 1: Introduction to Astronomy • Pages 134–135			
	Day 57	Do Lesson 2: The Earth Is Moving • Pages 136–139			
	Day 58	Do Lesson 3: Why Do We Have Seasons? • Pages 140–142			
	Day 59				
	Day 60				

Date	Day	Assignment	Due Date	✓	Grade
Week 4	Day 61	Do Lesson 4: Telescopes • Pages 143–145			
	Day 62	Complete **Space Models and Tools Unit Vocabulary Review** (Lessons 1–4) • Page 146			
	Day 63	Universe for Beginners Unit 2: Outer Space Do Lesson 5: Overview of the Universe • Pages 147–150			
	Day 64				
	Day 65				
Week 5	Day 66	Do Lesson 6: Stars • Pages 151–153			
	Day 67	Do Lesson 7: Our Galaxy • Pages 154–156			
	Day 68	Do Lesson 8: Asteroids • Pages 157–159			
	Day 69				
	Day 70				
Week 6	Day 71	Do Lesson 9: Comets • Pages 160–162			
	Day 72	Do Lesson 10: Meteors • Pages 163–165			
	Day 73	Complete **Outer Space Unit Vocabulary Review** (Lessons 5–10) • Page 166			
	Day 74				
	Day 75				
Week 7	Day 76	Universe for Beginners Unit 3: Sun and Moon Do Lesson 11: Our Solar System • Pages 168–170			
	Day 77	Do Lesson 12: Our Sun • Pages 171–173			
	Day 78	Do Lesson 13: The Surface of the Sun • Pages 174–176			
	Day 79				
	Day 80				
Week 8	Day 81	Do Lesson 14: Solar Eclipse • Pages 177–180			
	Day 82	Do Lesson 15: Solar Energy • Pages 181–182			
	Day 83	Do Lesson 16: Our Moon • Pages 183–185			
	Day 84				
	Day 85				
Week 9	Day 86	Do Lesson 17: Phases of the Moon • Pages 186–187			
	Day 87	Do Lesson 18: Where Did the Moon Come From? • Pages 188–189			
	Day 88	Complete **Sun and Moon Unit Vocabulary Review** (Lessons 11–18) • Page 190			
	Day 89				
	Day 90				
		Mid-Term Grade			

Date	Day	Assignment	Due Date	✓	Grade
		Second Semester-Third Quarter			
Week 1	Day 91	Universe for Beginners Unit 4: Planets Do Lesson 19: Mercury • Pages 192–194			
	Day 92	Do Lesson 20: Venus • Pages 195–197			
	Day 93	Do Lesson 21: Earth • Pages 198–201			
	Day 94				
	Day 95				
Week 2	Day 96	Do Lesson 22: Mars • Pages 202–205			
	Day 97	Do Lesson 23: Jupiter • Pages 206–208			
	Day 98	Do Lesson 24: Saturn • Pages 209–210			
	Day 99				
	Day 100				
Week 3	Day 101	Do Lesson 25: Uranus • Pages 211–213			
	Day 102	Do Lesson 26: Neptune • Pages 214–215			
	Day 103	Do Lesson 27: Pluto and Eris • Pages 216–217			
	Day 104				
	Day 105				
Week 4	Day 106	Complete **Planets Unit Vocabulary Review** (Lessons 19–27) • Page 218			
	Day 107	Universe for Beginners Unit 5: Space Program Do Lesson 28: NASA • Pages 220–222			
	Day 108	Do Lesson 29: Space Exploration • Pages 223–225			
	Day 109				
	Day 110				
Week 5	Day 111	Do Lesson 30: Apollo Program • Pages 226–229			
	Day 112	Do Lesson 31: The Space Shuttle • Pages 230–232			
	Day 113	Do Lesson 32: International Space Station • Pages 233–235			
	Day 114				
	Day 115				
Week 6	Day 116	Do Lesson 33: Astronauts • Pages 236–239			
	Day 117	Do Lesson 34: Solar System Project • Page 240			
	Day 118	Complete your Special Project			
	Day 119				
	Day 120				
Week 7	Day 121	Do Lesson 35: Conclusion • Page 241–243			
	Day 122	Complete **Space Program Unit Vocabulary Review** (Lessons 28–35) • Page 244			
	Day 123				
	Day 124				
	Day 125				

Date	Day	Assignment	Due Date	✓	Grade
Week 8	Day 126	Planet Earth for Beginners Unit 1: Origins and Glaciers Do Lesson 1: Introduction to Earth Science • Pages 248–250			
	Day 127	Do Lesson 2: The Earth Is Special • Pages 251–253			
	Day 128	Do Lesson 3: The Earth's History • Pages 254–257			
	Day 129				
	Day 130				
Week 9	Day 131	Do Lesson 4: The Genesis Flood • Pages 258–260			
	Day 132	Do Lesson 5: The Great Ice Age • Pages 261–263			
	Day 133	Do Lesson 6: Glaciers • Pages 264–266			
	Day 134				
	Day 135				
Second Semester-Fourth Quarter					
Week 1	Day 136	Do Lesson 7: Movement of Glaciers • Pages 267–269			
	Day 137	Do the **Origins and Glaciers Unit Vocabulary Review** (Lessons 1–7) • Page 270			
	Day 138	Planet Earth for Beginners Unit 2: Rocks and Minerals Do Lesson 8: Design of the Earth • Pages 272–274			
	Day 139				
	Day 140				
Week 2	Day 141	Do Lesson 9: Rocks • Pages 275–277			
	Day 142	Do Lesson 10: Igneous Rocks • Pages 278–279			
	Day 143	Do Lesson 11: Sedimentary Rocks • Pages 280–282			
	Day 144				
	Day 145				
Week 3	Day 146	Do Lesson 12: Fossils • Pages 283–285			
	Day 147	Do Lesson 13: Fossil Fuels • Pages 286–288			
	Day 148	Do Lesson 14: Metamorphic Rocks • Pages 289–291			
	Day 149	Do Lesson 15: Minerals • Pages 292–293			
	Day 150				
Week 4	Day 151	Do Lesson 16: Identifying Rocks • Pages 294–295			
	Day 152	Do Lesson 17: Rock Cycle • Pages 296–298			
	Day 153	Do Lesson 18: Gems • Pages 299–301			
	Day 154	Complete **Rock and Minerals Unit Vocabulary Review** (Lessons 8–18) • Page 302			
	Day 155				
Week 5	Day 156	Planet Earth for Beginners Unit 3: Mountains and Movement Do Lesson 19: The Earth Has Plates • Pages 304–307			
	Day 157	Do Lesson 20: Mountains • Pages 308–309			
	Day 158	Do Lesson 21: Types of Mountains • Pages 310–312			
	Day 159	Do Lesson 22: Earthquakes • Pages 313–315			
	Day 160				

Date	Day	Assignment	Due Date	✓	Grade
Week 6	Day 161	Do Lesson 23: Preparing for an Emergency • Pages 316–318			
	Day 162	Do Lesson 24: Volcanoes • Pages 319–321			
	Day 163	Do Lesson 25: Volcano Types • Pages 322–324			
	Day 164	Do Lesson 26: Mount St. Helens • Pages 325–327			
	Day 165				
Week 7	Day 166	Complete **Mountains and Movement Unit Vocabulary Review** (Lessons 19–26) • Page 328			
	Day 167	Planet Earth for Beginners Unit 4: Water and Erosion Do Lesson 27: Geysers • Pages 330–332			
	Day 168	Do Lesson 28: Erosion • Pages 333–335			
	Day 169				
	Day 170				
Week 8	Day 171	Do Lesson 29: Landslides • Pages 336–338			
	Day 172	Do Lesson 30: Stream Erosion • Pages 339–341			
	Day 173	Do Lesson 31: Soil • Pages 342–343			
	Day 174	Do Lesson 32: Grand Canyon • Pages 344–346			
	Day 175				
Week 9	Day 176	Do Lesson 33: Caves • Pages 347–349			
	Day 177	Do Lesson 34: Rock Collection • Pages 350–351			
	Day 178	Do Lesson 35: Appreciating Planet Earth • Pages 352–353			
	Day 179	Complete **Water and Erosion Unit Vocabulary Review** (Lessons 27–35) • Page 354			
	Day 180				
		Final Grade			

Weather and Water
for Beginners

Atmosphere and Meteorology

Weather and Water
for Beginners

GOD'S DESIGN®

Lessons 1-4

God Made Weather

What is the weather like outside right now? God created all different kinds of weather. What kinds of weather can you think of? In most parts of the world, the weather is different during different times of the year.

We call these different times of the year seasons. Do you know what the four seasons are called? They are spring, summer, fall (or autumn), and winter.

In the spring, the weather is often cool and rainy at the beginning but becomes warmer later on. Weather is usually warm and sunny in the summer. In the fall, the weather starts to cool down and leaves usually fall off of trees. Winter is when the weather is the coldest. Many places get snow in the winter.

When the world was first created, God said that everything was very good. But Adam and Eve sinned and disobeyed God. As part of their punishment God cursed the earth. This resulted in different weather than before. Later there was a flood over the whole world. This made even more changes to the weather. So the weather we have today is different from the weather at creation.

Sometimes weather events like storms can be very dangerous. But God is in charge. So you do not need to be afraid of the weather. The Bible tells us that God controls the weather. Jesus spoke to the storm, and the wind and the rain stopped (Matthew 8:23–27). We can be thankful that God is in control.

? **What do you like to do on a sunny day?**

? **What kinds of things can you do on a rainy day?**

? **What are your favorite things to do on a snowy day?**

? **What are the four seasons?**

Scripture Trace

What sort of man is this, that even winds and sea obey him?   Matthew 8:27

Favorite Season

Fill in the missing vowels of the names of the four seasons. Then draw a line to match each picture with the right season. Circle your favorite season.

s _ mm _ r

a _ t _ mn

spr _ ng

w _ nt _ r

The Atmosphere

Earth is a very special planet. It is the only planet that has people and animals living on it. Earth is surrounded by air. Another name for air is the **atmosphere**. The earth's atmosphere is made up of mostly **nitrogen** and **oxygen**. Animals and people need oxygen to breathe. No other planet has the right amount of oxygen in its atmosphere.

Air protects us from the very hot and very cold temperatures in space. Planets that do not have an

Scripture Trace

When it is evening, you say, "It will be fair weather, for the sky is red." — Matthew 16:2

atmosphere are very hot when the sun shines on them. They are very cold when the sun does not shine on them. Earth is not too hot or too cold.

The atmosphere is where weather happens. The air moves around. When it is windy, the air is moving quickly. When air moves, it moves rain and snow from one place to another.

The atmosphere also protects us from small meteors and other things in space that might crash into our planet. These space rocks usually burn up in the atmosphere before hitting the earth. We can see how important our atmosphere is by looking at the moon. The moon does not have an atmosphere. It is very hot in the sun and cold in the dark. It is covered with pits called craters where rocks have crashed into it. It does not have any life.

Air Has Oxygen

Place a lump of modeling clay onto a table. Push a small candle into the clay to hold it upright. Light the candle and watch it burn for a few seconds. The candle can burn because there is oxygen in the air. Cover the candle with a glass jar. What happened to the flame? It goes out after a few seconds. The flame uses up the oxygen in the air. The jar stops more air from getting to the candle. How long does it take for the flame to go out if you use a bigger jar?

? What is the atmosphere?

? What are the two main ingredients in air?

? What are some good things that our atmosphere does for us?

? Why does the moon have so many more craters than the earth?

Atmosphere Makes a Difference

Trace the land on the earth and color the land green and the water blue.
Trace the craters on the moon, then draw more craters.

The Weight of Air

How much does air weigh? Many people think that air does not weigh anything. Air is very light. But air actually does weigh a small amount. Air is made up mostly of oxygen and nitrogen. And these things have weight.

Gravity pulls down on everything on Earth. It even pulls down on the air. Air is pressing down on everything.

This is called __air pressure__.
Air pressure is important because it affects the weather.

Air Has Weight

You can see that air has weight by doing the following experiment.

1. Tape an empty balloon to each end of a yardstick.

2. Tie a string to the center of the yardstick. Tape the other end of the string to the edge of a table. The stick should hang down in front of the table. Adjust the string on the stick until you get the stick to balance. Then tape the string in place on the stick.

3. Remove one of the balloons. Fill it with air and tie it shut. Tape it back in the same place on the stick.

4. Watch what happens to the yardstick.

The side of the yardstick with the filled balloon will tip toward the ground. This is because the balloon with air weighs more than the balloon without any air. Air has weight.

Air Helps Things Fly

Many things are able to fly because of air pressure.
Count the number of things in the picture that can fly.
Color the things you found.

Write the number here:

? Does air have weight?

? What do we call air pressing down on things?

? Why is air pressure important?

Weather is something that affects everyone. People want to know if the weather will be nice enough for a picnic. Or they might want to know if it is going to snow tomorrow. Some people are so interested in weather that they study it as their job. Someone who studies the weather is called a

meteorologist.

Meteorologists look at what the weather is like today. They look at how the air is moving. Then they predict what the weather will be like tomorrow.

FRI 57 80

SAT 58 81

SUN

? What is a meteorologist?

? What are two things a meteorologist looks at to predict the weather?

Weather Wreath

Make a weather wreath. Cut out the center of a paper plate. Use the outer ring of the plate to make your wreath.

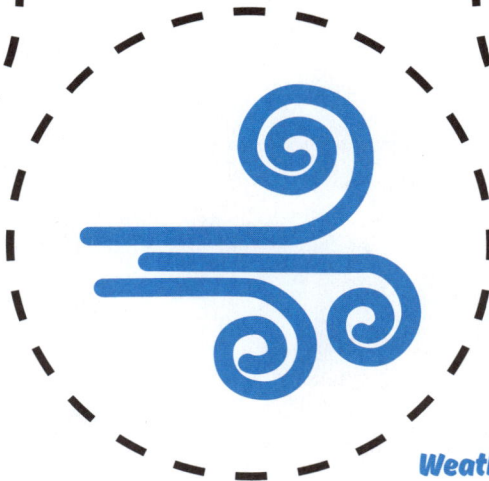

Cut out the weather symbols on this page. Color each of the symbols and glue them onto the wreath.

Hang the wreath where you can see it. This can remind you of the different kinds of weather you might have.

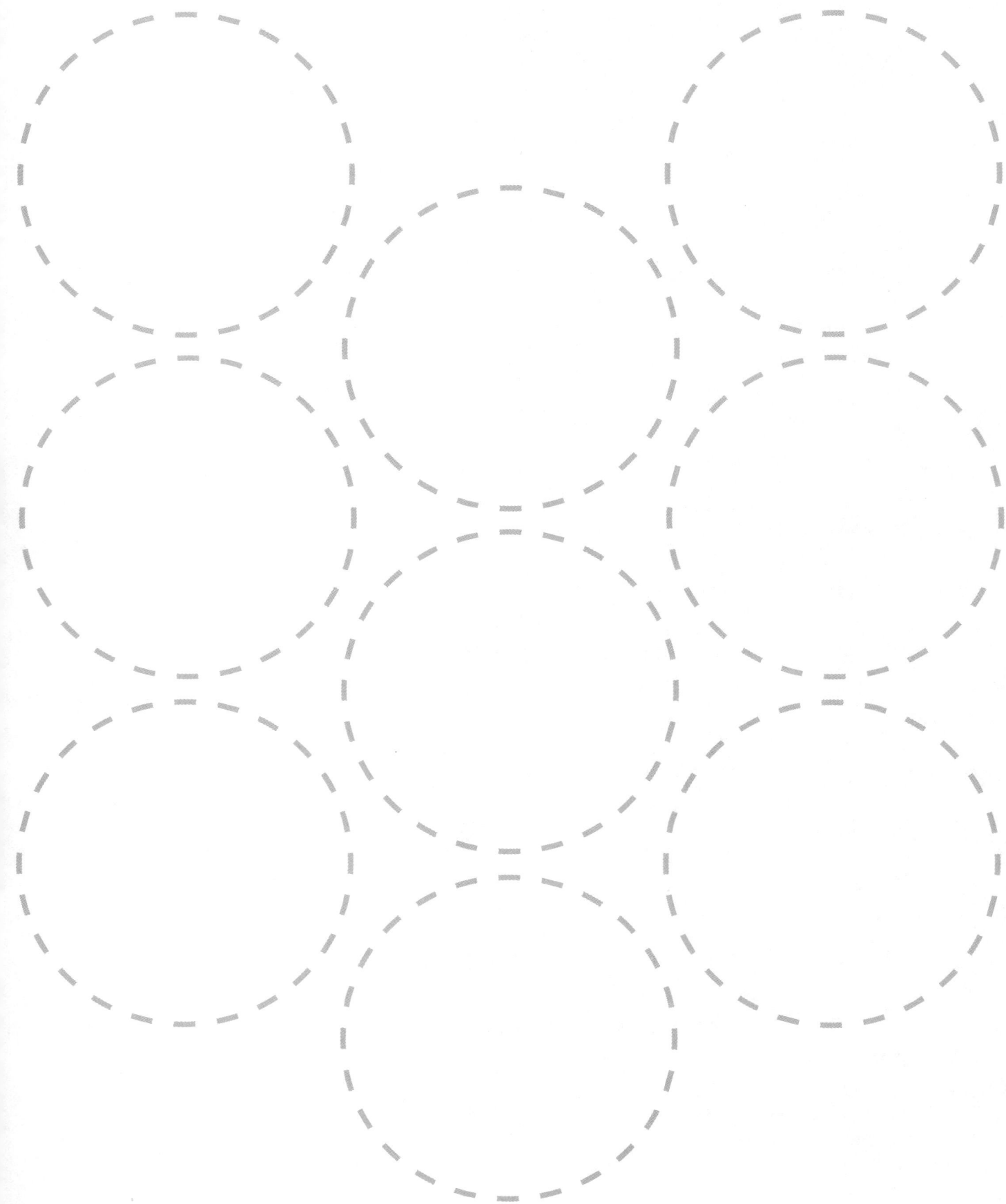

Predict the Weather

Circle the weather picture that should come next in the pattern.

Unit Vocabulary Review

Draw a string from the definition balloon to the correct vocabulary word gift.

Air surrounding the planet

Main ingredients in air

Air pressing on everything

Person who studies weather

Telling what the weather should be like tomorrow

nitrogen and oxygen

meteorologist

predict

atmosphere

air pressure

Ancient Weather and Climate

Weather and Water
for Beginners

Lessons 5-8

GOD'S DESIGN®

Weather vs. Climate

Sometimes the weather is sunny and warm. Other times the weather is rainy and cold. Weather changes from day to day.

Weather is what is happening outside right now.

Another word that is similar to weather is climate. But climate does not mean the weather right now. *Climate* tells you what the weather is like throughout a whole year. Different parts of the world have different climates.

The North Pole and South Pole are cold and snowy during most of the year. They have a *polar* climate.

Areas with a *desert* climate are very dry. Deserts are usually hot, but not always.

Tropical climates are warm. Tropical areas have lots of rain.

Most of the United States has a

temperate

climate. A temperate climate is warm in the summer and cold in the winter.

The weather will change from day to day. But the climate will remain about the same from year to year.

Climates of the World Booklet

Make a climate booklet showing the different climates you learned about in this lesson. Collect pictures from magazines (or print out from the internet with your parent's permission) of places in the four climates. Glue the pictures onto construction paper pages. Make one or more pages for each climate. Make a title page for your book. Then punch holes in the pages and use yarn to tie the pages together.

? If I say that it is raining outside, am I talking about weather or climate?

? If I say that it is usually cold in the winter, am I talking about weather or climate?

? Name three different climates.

✏️ Climate Match Worksheet

Fill in the missing letters of the climate names. Then draw
a line to match each animal to the climate it lives in.

_ _ o _ i _ a

_ _ _ e _ e _ _

_ _ e _ e _ a _ e

_ o _ a _

Climate Before the Flood

The climate of the earth was different when the earth was first created. The Bible says that Adam and Eve did not need clothes in the garden of Eden. This means the climate was warm. Fossils of tropical plants have been found in nearly every part of the earth. So, we know that the weather was probably warmer around the whole world than it is today. Noah's Flood caused many changes to the world. This changed the weather and the climate of the earth.

? What clues does the Bible give about the climate before the Flood?

? What clues do fossils give us about the climate before the Flood?

Garden of Eden Worksheet

Find the hidden letters in the picture. Place them in the correct blanks to spell the likely type of climate in the Garden of Eden. Color the picture.

_ _ _ o _ i _ a _

The Great Flood

When Noah lived, the people on Earth were very wicked. They did not worship God. So, God decided to punish them. God told Noah to build a giant boat called the ark. God sent animals to Noah. He put the animals in the ark. Noah and his family went into the ark, too.

Then God made it rain. It rained for forty days. People call this the _Great Flood_. There was so much water that the whole earth was covered with it. Everyone who was not on the ark died. This was God's punishment for people's sin. All of the land animals that were not on the ark died, too. But Noah and his family took care of the animals on the ark.

Rainy Days

Think about what it would be like to have 40 days of rain. Ask your parent for an old calendar. Mark off 40 days on the calendar. About how many weeks would that be?

After the Great Flood was over, God made a promise. He promised he would never flood the entire earth again. As a symbol of that promise, God made a rainbow in the sky with the colors of red, orange, yellow, green, blue, indigo, and violet.

? Why did God send the Great Flood?

? Who was killed by the Flood?

? Who was saved from the Flood?

? What promise did God make when he sent the rainbow?

? What are the colors of the rainbow?

Scripture Trace

I have set my bow in the cloud.   Genesis 9:13

Rainbow Worksheet

Trace the color words, then color the rainbow. Be sure to place the colors on the rainbow in the order they should go. To remember what order the colors appear in a rainbow, you can use the letters in the name Roy G. Biv.

R O Y G B I V

■ blue   ■ yellow

■ violet   ■ red   ■ green

■ orange   ■ indigo

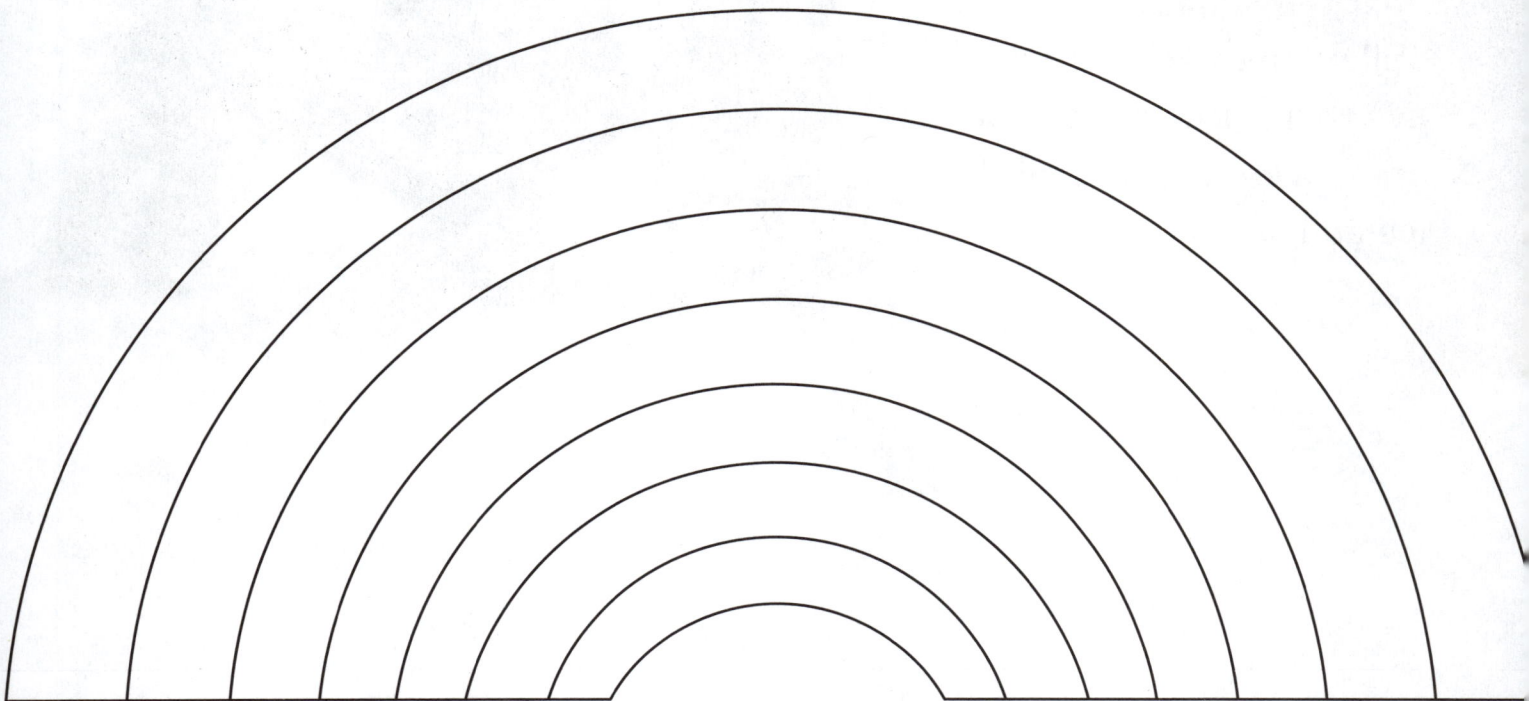

Climate After the Flood

After the Flood, the climate was different. The temperatures around the world were cooler because there were many clouds. In some areas, there was a lot of snowfall. This caused what is called the

Ice Age.

During the Ice Age, areas near the equator were still warm. People could live there. But many other areas were covered with ice and snow.

✏️ Scripture Trace

Never again shall there be a flood to destroy the earth.

Genesis 9:11b

After a while, temperatures warmed up and melted much of the ice. As the ice melted, it flowed into the oceans. But not all of the ice has melted. Some of it remains near the North and South Poles in the form of glaciers. After hundreds of years, the weather became very much like it is today. We can thank God that even during the Ice Age, he provided a way for people to live.

Mr. Sunshine

Today we do not live in an ice age. We have many warm, sunny days. You can make a fun picture of the sun by doing the following.

1. Place a cereal bowl in the center of a piece of paper and trace around it to make the center of your sun.

2. Color the circle yellow and draw a face in the middle.

3. Dip your hand in yellow finger paint and make a handprint at the edge of the circle. Repeat making hand prints all around the circle. These are the rays of sunshine coming from the sun.

? Why was there an ice age after the Flood?

? What happened to all the ice from the Ice Age?

Unit Vocabulary Review

Write the missing words from the clues below into the puzzle.
Hint! There are letters to trace that can help you solve the puzzle.

Across

1. A _____ climate has warm summers and cold winters.
3. The great _____ was a punishment for people's sin.
5. A _____ climate is warm and rainy.
6. _____ is the weather over a long period of time.
8. A _____ is very dry and usually hot.

Down

2. A _____ climate is cold and snowy.
4. Rain and snow are two types of _____
7. The _____ Age had lots of snow and cold after the Flood.

Clouds

Weather and Water
for Beginners

Lessons 9–12

GOD'S
DESIGN®

Water Cycle

God has provided a way for plants, animals, and people to get the water we need. It is called the _water cycle_. Water from the oceans, lakes, and rivers is warmed by the sun. This warmth causes some of it to _evaporate_ into the air. Wind blows the wet air away from the water to the land. When the air cools, it forms clouds. This is called _condensation_. Then the water falls to the ground as rain or snow. The rain waters plants. People can use the water, too. Some rain flows into rivers or lakes. Finally, the rivers take the water back to the ocean. Then the cycle starts over again. Water is used over and over again so we never run out. Isn't God a great designer?

The Water Cycle

Fill in the blanks with the correct words.

1. When water is warmed by the _____,
   it evaporates. It _____ into the air.

2. _____ blows the air to the land.

3. _____ form. _____ or
   snow falls to the ground.

4. Rivers take water back to the _____.

clouds

ocean

rain

rises

sun

wind

? How does water get from the ocean into the air?

? How does water get from the sky to the land?

? How does water get from the land to the ocean?

? Why is the water cycle important?

Forming Clouds

It is fun to look for pictures in the clouds. It is also fun to learn about what clouds are and how they form. What do you think a cloud is? Does it look like a giant cotton ball or cotton candy? A

cloud is really a bunch of water drops in the sky.

Before there can be clouds, the sun has to shine. When the sun warms the water in oceans and lakes, some of the water turns into gas, or vapor. The water vapor goes into the air (evaporates). As long as the air is warm, the water will stay in its gas form. But when the air cools, the water turns back into its liquid form.

✏️ Scripture Trace

God covers the heavens with clouds. Psalm 147:8

Making a Cloud in a Jar

1. Place one inch of water in a jar. Seal it with a lid.

2. Place the jar in a pan of water. Heat the water over medium heat. This will heat the water inside the jar just like the sun heats the water in the ocean. Some of the water will turn to gas.

3. Place a plastic zip bag full of ice on top of the jar lid. This will cool down the gas at the top of the jar. Some of the gas will turn back into water drops. This will form a cloud at the top of the jar.

4. After a while some of the water drops may get big enough to fall back into the jar. This completes the water cycle.

Remember that this cooling is called condensation. These drops of water are very tiny. When there are many drops of water in the same area, they become a cloud.

Warm air rises. Rising air pushes up on the tiny drops of water. This helps to keep the clouds up in the sky and allows them to float. Sometimes, the water drops change back into gas and the cloud disappears. Other times, the cloud can last for many hours. And sometimes, the water falls to the ground as rain or snow.

? What happens to water when the sun shines on it?

? How can a cloud float?

Pictures in the Clouds Worksheet

Connect the water droplets to see an animal in the clouds. What other things do you see in the sky?

5   6   7   8
4   13
38   12 14
             9 15
37   3   10 11 16
36   2   1 17
34 35   19 18
33 32 31   20
29 30   24 21
28 27 26 25 23 22

Cloud Types

All clouds do not look the same. Sometimes the whole sky is covered with clouds. They look like a solid sheet across the sky.

These are called _stratus_ clouds.

Other clouds are big and fluffy. We call these

cumulus

clouds.

Stratus

Cumulus

And some clouds are wispy and curly. These kinds of clouds are called

cirrus clouds.

Look outside. Are there any clouds in the sky? See if you can tell what kind of clouds you are looking at. Use the pictures in this lesson to help you decide. Check back later in the day to see if the clouds have changed.

Cirrus

? How does a stratus cloud look?
? How does a cumulus cloud look?
? How does a cirrus cloud look?

Scripture Trace

For your steadfast love is great to the heavens, your faithfulness to the clouds.

Psalm 57:10

name _____

Cloud Picture

Use cotton balls to make a picture with different types of clouds. Glue the cotton balls to a piece of blue construction paper. Leave some cotton balls round and fluffy like cumulus clouds. Stretch some out into long strips like stratus clouds. Tear others into small pieces and twist and curl them around like cirrus clouds. You can add other things to your picture like trees, animals, and people, too.

Precipitation

Clouds form when water in the air cools and turns into liquid. These water drops are very tiny. Sometimes these tiny drops stick together. They form bigger drops. When the drops get big enough, they become heavy. The rising air cannot keep them in the sky. Then they fall to the ground. Water falling from clouds is called

precipitation.

The most common type of precipitation is rain.

During the winter, clouds can be very cold. The water in the cloud may freeze. It can form ice crystals. Then the water

becomes snow. All snowflakes have six sides. But every snowflake has its very own shape. No two are exactly alike.

During the summer, the bottom of a cloud may be warm. But the top of the cloud may be cold. The warm air can push water drops from the bottom of the cloud up to the top. The water drops freeze in the top of the cloud. Then they fall back down to the bottom of the cloud. There, more water droplets stick to the frozen water. They can then be pushed back up to the top of the cloud and freeze again. This can happen over and over. Eventually, the balls of ice get too heavy to be pushed back up. They fall to the ground. These ice balls are called

hail.

Making Snowflakes

Start with a square piece of white paper. Fold it in half diagonally. Fold the triangle in half again. Now fold the triangle into thirds. Cut across the bottom to remove the "tails". Cut out pieces of the triangle around the edges. Experiment cutting different sizes and shapes. Open the triangle to reveal a unique snowflake.

? What is precipitation?

? Name three types of precipitation.

? How is hail formed?

Scripture Trace

For he . . . sends rain on the just and on the unjust.
Matthew 5:45b

Unit Vocabulary Review

Fill in the blank with the correct vocabulary word. The first letter of each word is given for you.

cirrus, cloud, condensation, cumulus, evaporates, float, hail, precipitation, rain, snow, stratus, water cycle

1. Water is reused through the W_____ C_____.

2. Water e_____ when it turns into a gas.

3. C_____ happens when water turns into a cloud.

4. A C_____ is formed when water turns into a liquid in the sky.

5. A cloud can f_____ because warm air pushes up on it.

6. S_____ clouds form a big sheet of clouds in the sky.

7. C_____ clouds are big and fluffy.

8. C_____ clouds are curly and light.

9. Water falling from the sky is called P_____.

10. R_____ is the most common form of precipitation.

11. S_____ forms when water turns into ice crystals in a cloud.

12. H_____ forms when water is pushed up into a cold part of a cloud over and over again.

Storms

Weather and Water
for Beginners

Lessons 13-17

GOD'S
DESIGN®

Air Masses and Weather Fronts

Weather takes place in the air around us. A large amount of air that is all the same temperature is called an __air mass__. Air masses move from place to place around the world. The air outside your house may have been in another state yesterday. A breeze will move the air mass from one place to another. That is when weather can get interesting.

A bunch of warm air can bump into a bunch of cold air. Where these two air masses touch is called a __weather front__. The air gets all jumbled up at the front. Some of the warm air cools down. Some of the cold air warms up. This mixing of air usually causes clouds to form. Then there can be rain or snow, and it often becomes windy. Meteorologists track the movement of air masses and weather fronts so they can predict what the weather will be like in the next few days.

Weather Songs

Sing some fun weather songs together. Sing "You Are My Sunshine" and "The Itsy Bitsy Spider." Then sing the following words to the tune of "If You're Happy and You Know It."

If you see the leaves moving on the trees,
If you see the leaves moving on the trees,
If you see the leaves moving then the wind is clearly blowing.
If you see the leaves moving on the trees.

If you hear thunder booming cover your ears.
If you hear thunder booming cover your ears.
If you hear thunder booming and see lightning that is zooming,
If you hear thunder booming cover your ears.

Now make up your own verses to this song:

_____

_____

_____

_____

_____

_____

_____

? What is an air mass?

? What is a weather front?

? What happens at a weather front?

Wind

Wind is air that is moving quickly. Wind is a very important part of weather. Wind moves air masses from one place to another, which can bring rain or snow. But what causes the wind to blow?

When the sun shines it warms the land, water, and air. But land warms up faster than water. So, air over land is warmer than air over the water. Warm air is more spread out than cold air. So cold air moves into areas that have warm air. This moving air is wind. The wind is always blowing somewhere. God designed our world so that it would be warmed by the sun. And this would cause the wind to blow. Wind brings the weather that gives us rain. God sure takes care of us!

Make a Wind Sock

You can make a wind sock. It will show what direction the wind is blowing.

Pull down on the center of a metal clothes hanger. This will bend the hanger so that it is shaped like a square or diamond.

Tape the opening of a trash bag to the metal hanger frame.

Take your wind sock outside. Hold it out by the hook. Slowly turn until the wind begins to fill the trash bag. This shows what direction the wind is blowing.

? **What is wind?**

? **What causes the wind to blow?**

? **Why is wind important?**

Scripture Trace

The wind blows where it wishes, and you hear its sound.   —John 3:8

Using Wind Worksheet

Wind is useful for moving weather fronts. But it is useful for many other things as well. Circle everything in the picture that uses wind.

Thunderstorms

Thunderstorms are huge rainstorms with big clouds, strong winds, and lots of rain. The lightning and thunder that come with these storms can be scary to some people. But they are just a natural part of the weather that God created. Thunderstorms are very important. They bring much needed rain to many areas of the world.

Thunderstorms happen when the air is warm and there are a lot of water droplets in the air. In temperate areas, these storms usually form in the summer. They can happen anytime of the year in tropical areas. There are about 50,000 thunderstorms around the world each day. Most of these thunderstorms happen near the equator, in areas around the middle part of the earth.

There are strong winds inside a thunderstorm. The winds blow the water drops against each other. This causes electricity to build up inside the cloud. After a while, the built-up electricity flows through the cloud and there is a huge flash of light. We call this

lightning. Lightning is very hot. It heats the air around it to a very high temperature. This hot air moves outward very quickly and makes the loud sound that we call

thunder.

Although lightning and thunder may seem scary, they are not harmful if you are inside a building. You definitely don't want to stay outside during a thunderstorm. Enjoy watching the storm safely inside your home to avoid being hurt.

Making Lightning

You can safely make mini-lightning inside your own home. All you need are a fuzzy stuffed animal and a piece of cloth. Go into a dark room. Rub the stuffed animal with the cloth several times. Slowly pull the cloth away. You should see lots of little sparks jump between the cloth and the animal. These sparks are just like the lightning jumping from the cloud to the ground during a storm, but they are not dangerous.

? Are thunderstorms more likely to happen in the summer or the winter?

? What causes lightning?

? What causes thunder?

? Where should you be during a thunderstorm?

Thunderstorm

Connect the dots to see lightning coming out of the storm cloud.
Then, color the picture.

A
Q
1
19
B
P
2
18
C
O
3
17
D
N
4
16
E
M
5
15
F
L
6
14
G
K
7
13
H
J
8
12
I
9
11
10

Tornadoes

tornado

Tornadoes are dangerous storms. A *tornado* is a swirling cloud that acts like a giant vacuum cleaner. It sucks up anything in its path. Tornadoes can form at the edge of a thunderstorm. Warm air is moving upward, and cool air is moving downward. This causes a swirling motion in the air. If there is enough heat in the area, this swirling air can turn into a tornado. Most tornadoes occur in the spring in the central and eastern parts of the United States.

Tornadoes usually do not last very long. But they can cause a lot of damage by knocking down trees and buildings and by blowing things around. You need to pay attention to weather warnings. If there is a tornado warning for your area, you need to go to a safe place. It is safest in a basement, under a set of stairs, under a heavy table, or in a bathtub with cushions on top of you. Staying in a safe place will help protect you if a tornado comes through your neighborhood.

Tornado in a Bottle

You can see a tornado in a bottle. Fill an empty 2-liter soda bottle half full of water. Place a second bottle upside down on top of the first so the mouths are touching. Use a plastic tornado tube or duct tape to connect the bottles together. Hold the bottles over a sink. Turn them upside down so the water is in the top. Swirl the top bottle in a circle for a few seconds. The water should form a swirling tornado near the mouth of the bottle.

? What is a tornado?

? When do most tornadoes occur?

? Where do most tornadoes occur?

? Where is the safest place to be during a tornado?

Tornado Alley Map

So many tornadoes occur in the central United States that this area is called Tornado Alley. Color the area in the center of the map red. Do you live in Tornado Alley?

Hurricanes

Hurricanes are the largest and fiercest storms in the world.

Hurricanes are storms that form over warm water in the ocean. Almost all hurricanes form near the equator. Most hurricanes form in the Pacific Ocean near Indonesia and the Philippines. Others form in the Atlantic Ocean near Senegal and Mauritania. You can look at a globe to see where these countries are located. Hurricanes are also called typhoons in some parts of the world.

A hurricane starts out as a small thunderstorm over the ocean. Sometimes several thunderstorms form close to each other. If these storms move together, they can become one big storm. Sometimes this big storm continues to grow. If it stays in a warm location for a long time, it can become a hurricane.

Hurricanes spin very quickly. The winds inside a hurricane can be 100 miles per hour (45 meters per second). If the storm moves toward land, it can do a great deal of damage. If you and your family are in an area where a hurricane is expected to pass through, it is important to be prepared and follow any warnings from weather officials. It is not safe to stay in the path of a hurricane, and if told to, everyone should leave and go to a safe place.

Read a Weather Book

Visit your local library and see what books you can find about the weather. You may even find books that focus on specific weather topics, like tornadoes, blizzards, or other historic weather events.

? What are the largest storms in the world called?

? Where do hurricanes usually form?

? What is the center of a hurricane called?

Hurricane Worksheet

The very center of a hurricane is called the eye. There is very little wind inside the eye, but the winds are strong in the rest of the storm. Put an **X** on the eye of the hurricane.

✏️ Unit Vocabulary Review

Unscramble the letters to form vocabulary words. The first letter of each word is in **red**.

IRA SMAS _____

WHTAREE TORFN _____

DINW _____

HTUNDREMORST _____

GHILTNNIG _____

UNTHDER _____

TADORNO _____

URRHIENCA _____

Weather Information

Weather and Water
for Beginners

Lessons 18-22

GOD'S
DESIGN®

Measuring Temperature and Air Pressure

Do you remember what a meteorologist does? A meteorologist is a scientist who studies the weather. These scientists use many different instruments to measure the weather. They use a

thermometer to measure temperature. *Temperature* is how hot or how cold something is. Scientists measure the temperature of the air near the ground. They also measure the temperature of the air high in the sky. They do this by sending instruments up in the sky with balloons.

Measuring Temperatures

Use a thermometer to measure the temperature inside your house. Now measure the temperature outside. Is it warmer inside or outside? If it is a warm day outside, measure the temperature in the sun and in the shade. Which place had a lower temperature?

Meteorologists also need to measure air pressure. Recall that air pressure is how hard the air is pressing down on something. Scientists use an instrument called a __barometer__ to measure the air pressure. Air pressure changes when a storm front moves through an area. So scientists look for changes in air pressure to know where the weather fronts are. Knowing the temperature and air pressure helps meteorologists predict what the weather will be like tomorrow.

? What instrument do we use to measure temperature?

? What instrument do we use to measure air pressure?

? Why do meteorologists look for changes in air pressure?

Thermometer Worksheet

name _____

Thelma Thermometer measures the temperature. When it is hot, the liquid inside the thermometer goes up. When it is cold, the liquid goes down. Color the liquid inside Thelma red to show what the temperature is today. Barry Barometer measures the air pressure. Color the arrow to show what the air pressure is today.

Hint! If you have a barometer and a thermometer at your home, mark down the correct numbers. If you don't have them, you can check a weather app on a phone or watch the local weather to find the answers.

Thelma Thermometer   **Barry Barometer**

Measuring Rainfall and Wind Speed

Rain is an important part of weather. People often need to know how much rain falls during a storm. Scientists use an instrument called a _rain gauge_ to measure the rain. A rain gauge is just a cup with straight sides. The cup has marks to show how many inches of rain have fallen. The cup must be empty before the storm starts. When the storm is over, someone can look in the cup to see how much rain came down.

Scientists also like to measure how fast the wind is moving. This is called _wind speed_. They use an instrument called an _anemometer_ (an-uh-MOM-i-ter) to measure wind speed. An anemometer has little cups attached to a pole. When the wind blows it pushes the cups and makes them spin. The faster the wind blows, the faster the cups spin.

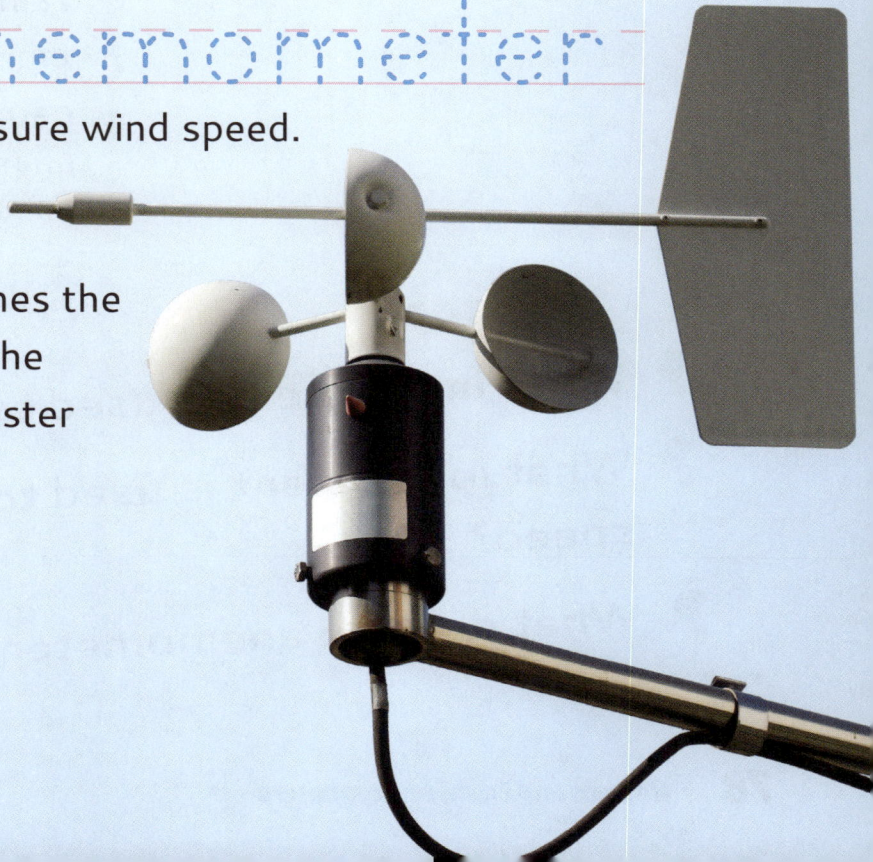

Make a Rain Gauge

Make your own rain gauge. Use a jar with straight sides. Put a strip of masking tape down one side. Use a ruler and a permanent marker to make 1–inch marks from the bottom of the jar. Place the jar in an open area in your yard. After it rains, go outside and see how much rain is in the jar. Be sure to empty the jar. Then your rain gauge will be ready for the next storm.

? What instrument is used to measure rainfall?

? What instrument is used to measure wind speed?

? What does an anemometer look like?

Recording Rain Worksheet

Mary and John used a rain gauge to measure rainfall each day this week. Draw a line on each gauge to show the correct water level for each day.

Monday – 1 inch
Wednesday – 2 inches
Friday – 2 1/2 inches

Tuesday – 1/2 inch
Thursday – 1 1/2 inches

On which day did they get the most rain? Circle that gauge. On which day did they get the least rain? Put a box around that gauge.

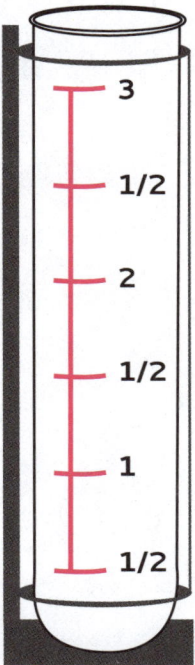
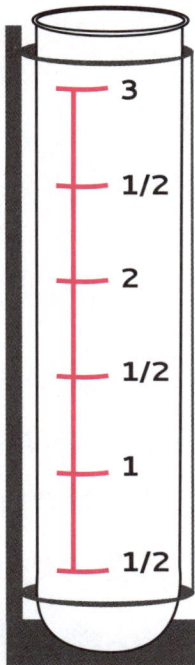
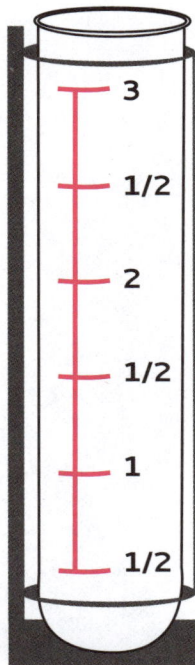
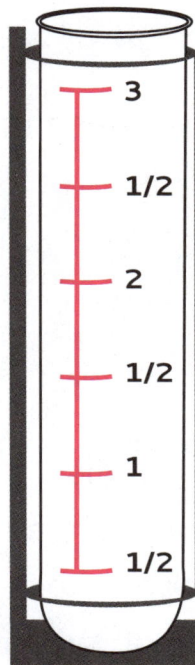
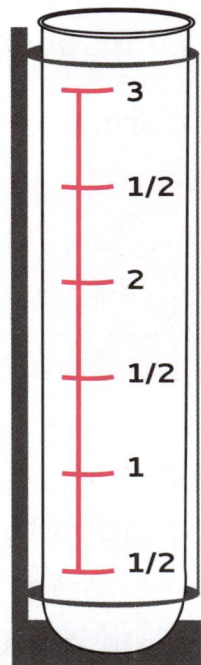

Monday Tuesday Wednesday Thursday Friday

Predicting Weather

Meteorologists measure temperature, air pressure, rainfall, and wind speed every day. They make many other measurements as well. People make these measurements all around the world. They even use balloons to send instruments up into the air. These instruments measure the temperature high in the atmosphere. Other instruments are on ships in the ocean, and some are on airplanes that fly through the atmosphere.

Information from these instruments is sent to the

National Weather Service.

Computers put all of this information together to make weather maps. These maps show what is going on with the weather all over the world. Meteorologists then use these maps to make predictions about what the weather will be like in the next few days.

? How do scientists get weather information out on the ocean?

? How do scientists get weather information from high in the sky?

? Where do meteorologists send all of their weather information?

? What do meteorologists do with their weather maps?

Weather Chart

Use this weather chart to keep track of the weather for the next few weeks. Each day draw a picture of what the weather was like. Draw a sun if the weather was sunny. Draw a snowflake if it snowed. Draw a raindrop if it rained. Draw a cloud if it was cloudy. Draw a kite if it was windy. If you want to practice being a meteorologist, you can predict what you think the weather will be tomorrow. Draw a small picture in the corner of the square for the next day, showing what you think the weather will be like. Then compare your prediction with what actually happened when you draw a larger picture of the actual weather.

Sunday	Monday	Tuesday	Wednesday	Thursday	Friday	Saturday

 Sun   Snow   Rain    Clouds    Wind

Weather Sayings

Many people have written sayings, poems, and nursery rhymes that help us better understand the weather. One famous saying is "March comes in like a lion and goes out like a lamb." What do you think this saying means? March often starts out windy because it is still winter. We say a strong wind roars like a lion. But spring begins toward the end of March, so it usually ends up being calm like a lamb.

Another saying uses a little rhyme: "April showers bring May flowers." This one tells us that it often rains in the springtime. Rain is an important part of helping the new plants grow. So even if it is a little cold and rainy early in the spring, we will be happy to see the flowers in May.

✏️ Scripture Trace

You know how to interpret the appearance of the sky.
Matthew 16:3

The Bible also talks about weather sayings. In Matthew 16:2–3, Jesus says, "When it is evening, you say, 'It will be fair weather, for the sky is red.' And in the morning, 'It will be stormy today, for the sky is red and threatening.' You know how to interpret the appearance of the sky, but you cannot interpret the signs of the times." This idea has been turned into a weather rhyme, "Red sky in morning, sailor take warning. Red sky at night, sailor's delight." This rhyme tells us that we can look at the color of the clouds and have an idea of what the weather will be like. All of these sayings can help you understand the weather a little better.

? What is your favorite weather saying?

Weather Nursery Rhyme Coloring Sheet

Color the pictures to help you remember the following weather nursery rhymes.

March comes in like a lion and goes out like a lamb.

April showers bring May flowers.

Rain, rain, go away, come again another day.

One misty moisty morning, when cloudy was the weather, I chanced to meet an old man clothed all in leather.

Weather Review

It is time for a quick review of what you have learned about weather. Weather begins with the sun. The sun shines on the earth and warms up the land, the water, and the air. Some places warm up faster than others. Some of the air ends up warmer than other air. Hot air rises and colder air falls. This causes the air to move around. The moving air is called wind.

The heat from the sun also causes water to evaporate from the ocean into the air. The moving air takes this water from place to place. And when the air cannot hold the large drops of water in it, the water falls to the ground as rain, snow, or hail.

What Will Ferin Wear?

Here is a fun game to help you think about different kinds of weather. The leader will describe a type of weather such as, "It is snowing." The leader will then say, "What will Ferin wear?" But instead of saying "Ferin," the leader will call out the name of one of the players. That player must say what he/she would wear in that kind of weather. The player should pretend to put that clothing on. That person then becomes the leader. The new leader chooses a different kind of weather and a new player to get dressed.

Scripture Trace

But let those who love Him be like the sun.
Judges 5:31 (NKJV)

Hot Air Rises

Changes in weather are caused by hot air rising and cool air falling. You can see hot air rising in the following experiment.

1. Stretch an empty balloon over the mouth of an empty glass bottle. We say the bottle is empty, but it is actually full of air.

2. Place the bottle in a pan with about an inch of water. Heat the water until it is boiling. The stove heats the water in the pan. It also heats the air in the bottle.

3. Watch the balloon. As the air in the bottle heats up, it rises and spreads out. The air fills the balloon, and the balloon gets bigger.

? Why is the sun so important for weather?

? What temperature of air rises?

? What temperature of air falls?

? What do we call moving air?

? What happens when the air gets so full of water that it cannot hold all of it?

Unit Vocabulary Review

Draw a line from each weather instrument to what it measures.
Then trace the words below each picture.

thermometer

air pressure

rain gauge

wind speed

barometer

temperature

anemometer

rain fall

What organization collects all of the weather data from
around the world?

N _____ W _____ S _____

Ocean Movements

Weather and Water
for Beginners

Lessons 23-29

Oceans

The oceans affect the weather. Most of the water that falls as rain and snow originally comes from the oceans. Many storms form over the oceans. So it is important to learn about the oceans.

There are five oceans around the world. The largest ocean is the

Pacific Ocean. The Pacific covers nearly one-third of the whole planet. The Pacific Ocean is on the west side of North America.

The second largest ocean is the Atlantic Ocean. The

Atlantic Ocean is on the east side of North America.

Name the Oceans

Memorize the names of the five oceans.

The _Indian Ocean_ is south of Asia.

The _Southern Ocean_ is located

near the South Pole. The _Arctic Ocean_

is near the North Pole. These oceans keep the water cycle going. This brings rain and snow to the whole earth. Look at a globe to see where each of these oceans is located.

? Which ocean is the biggest?

? Which ocean is on the east side of North America?

? Which ocean is near the South Pole?

? Can you name all five oceans without looking?

Oceans Worksheet

Locate each of the five oceans on this world map. Write the name of each ocean in the correct blank. Then color the oceans blue and the land brown.

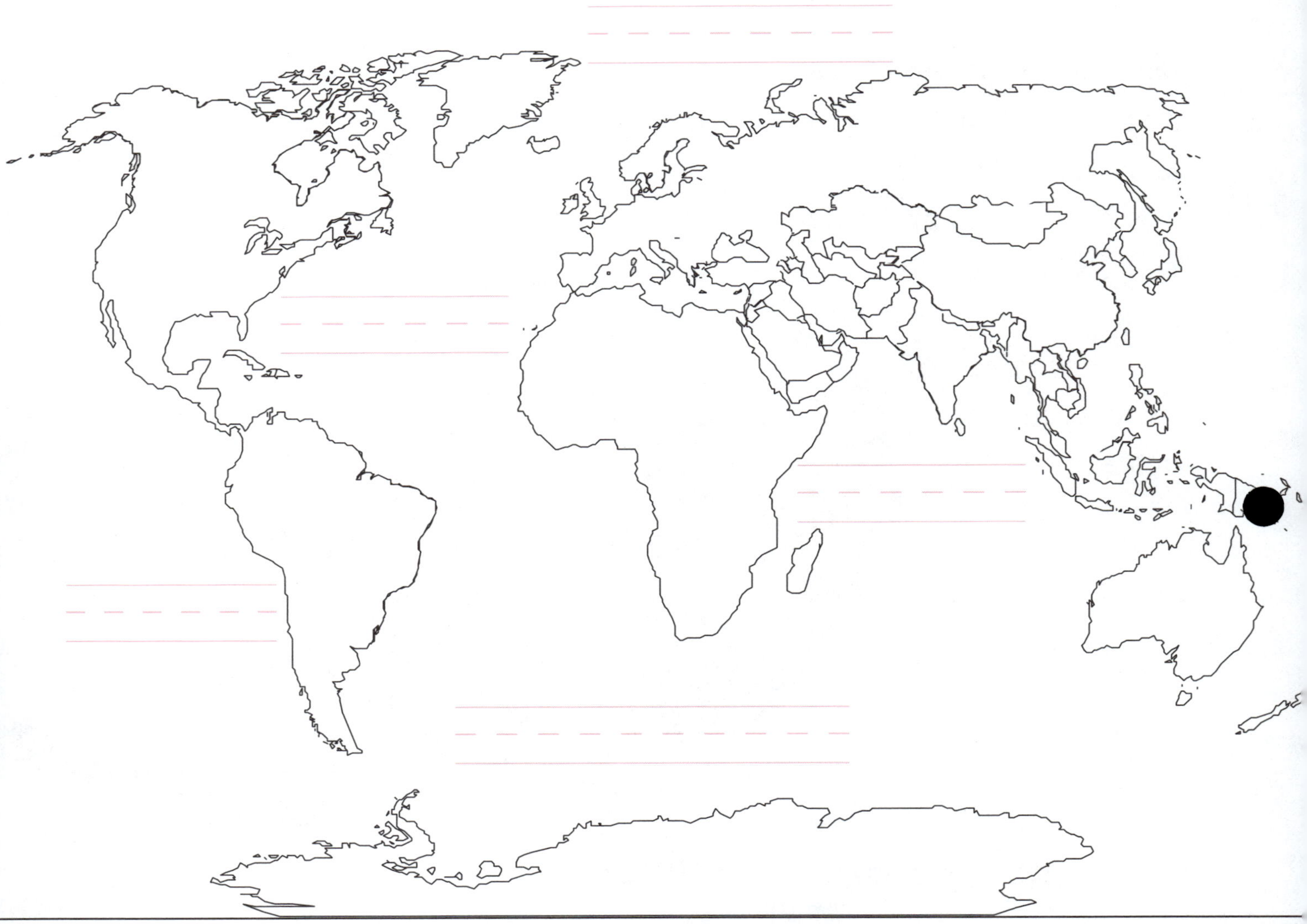

Arctic

Atlantic

Indian

Pacific

Southern

Why Is Seawater Salty?

If you have ever been to the ocean, you know that the seawater is salty. It is too salty to drink. It is too salty to water plants. Where did all that salt come from?

When it rains, water flows over the ground. There is a little bit of salt on the ground. A small amount of salt is

dissolved in or picked up by the water. This water flows into rivers, and rivers flow into the ocean. The amount of salt in the rivers is very small. We do not notice it.

Once the water gets to the oceans, it has nowhere to go. The only way water leaves the oceans is by evaporation, when the water goes into the air. When water evaporates, it leaves the salt behind. The amount of salt in the oceans builds up little by little over thousands of years. The plants, algae (seaweed), and animals that live in the ocean are specially designed by God to live in salty water.

Salt Pictures

Learn how salt is left behind in the oceans.

1. Heat a cup of water until very warm.

2. Stir in 2 teaspoons of salt. The salt seems to disappear when you stir it in, but it is still there. When the salt dissolves in the water, the water breaks it up into very tiny pieces that are too small to see. But you would still taste the salt if you tasted the water.

3. Use the saltwater to paint a picture on a sheet of dark colored construction paper.

Allow the picture to dry overnight. In the morning you should see a clear picture made of salt.

The water leaves the picture and moves into the air, just like water leaves the oceans and goes into the air. The salt was left behind on the picture just like the salt is left behind in the oceans.

Scripture Trace

Let your speech always be gracious, seasoned with salt.
Colossians 4:6

Ocean Currents

The water in the ocean is constantly moving. Let's find out why.

The sun shines more directly on the water near the

equator than it does near the North

and South Poles. So the water

near the equator is warmer. Winds blow the warm water away from
the equator. Cooler water moves away from the poles toward the
equator. This moving water makes little rivers in the ocean. We call

these rivers currents.

Arctic Ocean

Atlantic
Ocean

Pacific
Ocean

Indian
Ocean

Southern Ocean

Viewing Currents

The wind causes the water in the oceans to move. You can see how this works by doing the following experiment.

1. Fill a pie pan with water. Sprinkle pepper across the surface of the water.

2. Use a straw to blow gently across the surface of the water near one edge of the pan.

3. Watch how the pepper moves. It should look like little rivers have formed in the water. This is something like the way currents form in the ocean.

? Is the water warmer near the equator or near the North Pole?

? What direction does cooler water move?

? What do we call moving rivers in the ocean?

Ocean Currents

Color the arrows to show how the water moves in the oceans. Color the arrows that are pointing away from the equator red. This represents the warmer water moving away. Color the arrows that are pointing toward the equator blue. This represents the cooler water moving toward the equator. You can use the map earlier in the lesson as a guide.

Arctic Ocean

Atlantic Ocean

Pacific Ocean

Indian Ocean

Southern Ocean

Waves

Wind helps cause currents in the ocean. Wind also causes

_____ waves _____ in the ocean. When the wind blows, it picks up a little bit of water and moves it a short distance. The water then falls back into the ocean and pushes other water backward. The water that is moving up and down becomes a wave. Most waves start with wind pushing water.

Waves can move across the water for hundreds of miles. As a wave gets close to shore, the bottom of the wave touches the ocean floor. This causes the bottom of the wave to slow down. The faster water in the top of the wave builds up. The wave gets taller until it falls over.

These falling waves are called _____ breakers _____. After the wave breaks on shore, the water flows back out into the ocean.

Making Waves

You can make your own waves in a sink or bathtub. Fill the sink or tub with a few inches of water. Place a bottle on its side near the edge of the tub. Push the bottle into the water several times. Pushing water makes waves. Watch how the waves move across the bathtub.

Do this activity again using a larger bottle. The waves should be bigger because the bigger bottle pushes harder on the water. When a storm forms over the water, the strong winds act like the bigger bottle, pushing harder on the water and creating bigger waves.

? What causes a wave to form?

? Why do waves usually get taller near the shore?

? What do we call tall waves falling onto the shore?

Waves Coloring Sheet

Color this picture that shows how waves move toward the shore.

Tides

Have you ever built a sandcastle on the beach? Later in the day, as the ocean rose higher, the waves probably knocked it down. The level of the ocean is constantly changing. This regular rise and fall of the ocean is caused by the tides.

Tides are caused by the moon. How can the moon make the oceans move? The earth has gravity. Gravity pulls everything toward the earth. The moon also has gravity. The gravity from the moon is much less than the gravity from the earth since the moon is smaller. The moon's gravity pulls everything toward it a little bit. Even the oceans are pulled a little toward the moon. So as the moon moves around the earth, the ocean tries to follow it. This causes the level of the ocean along the seashore to change.

High tide is when the water comes far up on the shore. Low tide is when the water only comes a little way up the shore. There are two high tides and two low tides each day.

God designed the oceans to move. The moving water helps to keep the oceans clean. It also helps keep the temperatures more even around the world.

Follow the Moon Game

You will need a few friends or family members to help you play this game. Select one person to be the moon. Everyone else gets to be the ocean water.

Everyone except the moon stands facing each other in a circle. And then, they lock arms. The moon walks slowly around the outside of the ocean. As the moon gets close to someone they lean outward toward the moon. The person opposite this person should lean outward as well. The people on the sides lean in toward the center of the circle. This creates high and low tides. The tides will move around the circle as the moon moves around the circle.

After the moon has made one complete circle, that person chooses a new moon and takes their place in the circle.

? What causes the ocean level to change?

? How many high tides are there each day?

? Why did God design the ocean water to move?

Tides Worksheet

Which of these pictures shows the ocean at low tide? Which is at high tide?
Write the words and color the pictures.

High tide   **Low tide**

Wave Erosion

The movement of the ocean is helpful. It keeps the oceans clean and keeps the temperatures more even. But the movement of the water can be harmful, too. Waves crashing against the shore can wear away rocks and sand. When land is worn away, it is called

_____

erosion . Even tiny waves hitting the same place over and over can cause erosion.

When the water moves close to shore, it picks up bits of sand. The water then carries that sand out to deeper water. Waves also hit rocks on the shore. The water breaks off little bits of rock and carries them out to sea. So over time, the shore is worn away. During big storms like hurricanes, the shore can be eroded very quickly by the big waves.

Observing Erosion

To better understand how waves move sand away from the shore, do the following experiment:

1. Place an inch of sand in the bottom of a paint roller tray. Press more sand firmly on the ramp, as well.

2. Carefully pour water into the tray until it just touches the bottom of the ramp.

3. Place an empty plastic bottle in the deepest part of the water and gently move the bottle up and down in the water to create small waves.

Watch the waves move the sand down the ramp. Eventually most or even all of the sand will be moved into the deepest part of the tray. This is how sand is moved away from the shore by the ocean waves.

? **How does water wear away the shore?**

? **Do waves have to be big to cause erosion?**

Scripture Trace

The waters wear away the stones; the torrents wash away the soil of the earth. Job 14:19

Building Beaches

In the last lesson, you learned that erosion washes away bits of sand and rock from a beach. Fortunately, God designed ways for new sand and rock to takes its place. Waves can bring sand to a beach. Often, waves carry little bits of sand from deeper out in the ocean. When they come to shore, they drop this sand onto the shore. Waves also make new sand when they crash against rocks. They break off little bits of rock and break seashells which become part of the sand. Rivers can also break off little bits of rock as they flow toward the ocean. The river slows down as it reaches the ocean. Then these tiny bits of rock fall out of the water. They are added to the sand at the beach. So even though the waves move sand out into deeper water, they also bring more sand in. This helps to keep the shores where they are.

Sand in a Jar

Make a sand sculpture in a jar. Pour a layer of colored sand into a baby food jar. Make sure that it is not level. It should be higher in one area than another. Pour in another layer of a different color of sand. Continue adding different colors until the jar is completely full. Then screw on the lid. Do not shake your jar and your sculpture will last.

If you prefer, you can make your own "sand." Pour out a box of fruit ring cereal. Separate the rings into colors. Place one color of cereal in a plastic zip bag. Crush the cereal with a rolling pin. Repeat until all of the different colors have been crushed. Use this homemade sand to make your sand sculpture.

? How do waves create new sand?

? How do rivers create new sand?

Building with Sand Worksheet

Draw a picture of a sandcastle you would like to build if you were at the beach.

Unit Vocabulary Review

Label the following places on the world map. The first letter of each word is given to help you get started.

Atlantic Ocean   **North Pole**   **South Pole**

equator   **Pacific Ocean**

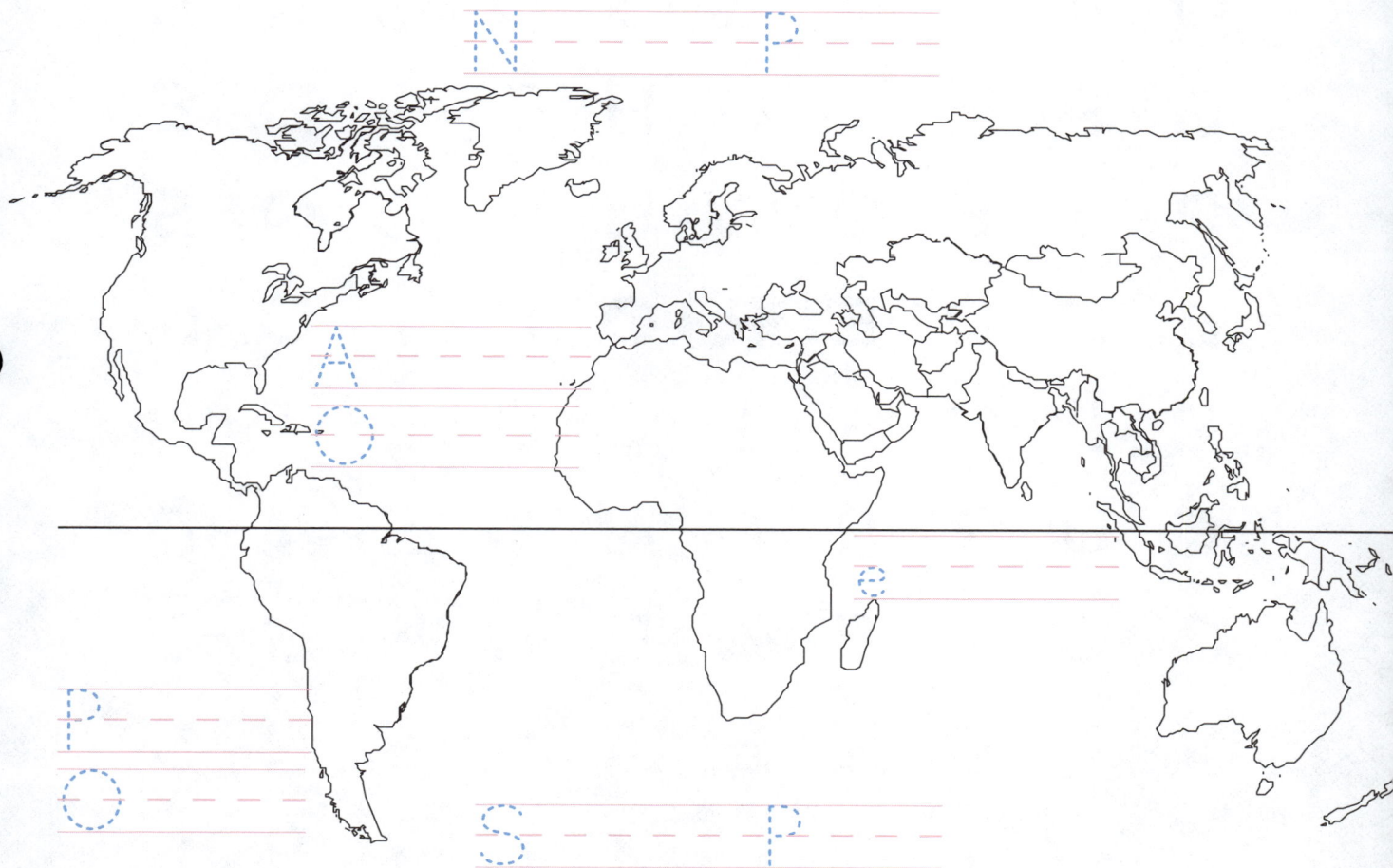

Unit Vocabulary Review

DISSOLVE   WAVES   HIGH TIDE   EROSION

CURRENTS   BREAKERS   LOW TIDE

```
Q W A V E S O M R R
G P H I G H T I D E
P J E Q W U G G Q S
B E O E R O S I O N
C L O W T I D E T H
C U R R E N T S Y D
O O L E W R B N U Z
V L B R E A K E R S
L W I C H Z L Y R L
I D D I S S O L V E
```

Seafloor

Weather and Water
for Beginners

Lessons 30-35

GOD'S
DESIGN®

Sea Exploration

One of the best ways to explore the ocean is to use a diving suit and air tank. An air tank and diving suit allow someone to stay underwater for about an hour. Divers can see many different kinds of fish, coral, seaweed, sharks, dolphins, and other animals. However, divers cannot go very deep in the ocean. The deeper you go the harder the water pushes down on you. So, divers cannot go too deep.

To explore deep in the ocean, people use small submarines called

submersibles. In a submersible, people can study what is near the bottom of the ocean. They can also use remote control cameras. They can take pictures of the ocean floor. Deep in the ocean people have seen sperm whales, glow-in-the-dark creatures, and living sponges. People have discovered that the ocean is filled with amazing living things that were designed by our amazing God.

Underwater Play

Sit under your kitchen table and pretend it is a deep-sea submarine. Pretend that you are an underwater explorer. Explore the ocean floor. What will you see there? If you take your submarine up closer to the surface, you can put on your diving suit and air tank. Then go into the water yourself to see the plants, algae, and animals up close.

? What can a diver wear to explore the ocean?

? What might a diver see in the ocean?

? Why can't divers go very deep into the ocean?

? How do people explore deeper in the ocean?

? What have people seen near the bottom of the ocean?

Sea Exploration Coloring Sheet

Color this picture showing different ways people explore the ocean.
Which living things can you name?

The Ocean Floor

When you look at the land around the world you see mountains, hills, valleys, and plains. When you look at the ocean, you see a fairly smooth surface compared to the land. What you can't see is the ocean floor. The floor of the ocean looks very much like the land. It has mountains, hills, valleys, and plains, too.

Imagine that you could take away the water and walk out into the ocean. You would see that the land slopes gently down from the shore. This area is called the **continental shelf**. Then suddenly the land drops steeply. This is called the **continental slope**.

Then the land levels out. This is called the **abyssal plain**. This is where you would find mountains and valleys under the water. It is very much like the land we see all around us. When you look at the ocean, remember there is much more below the surface.

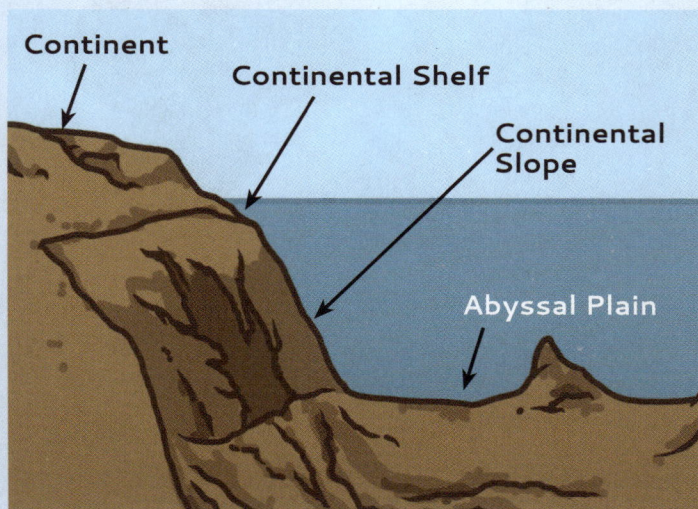

Continent

Continental Shelf

Continental Slope

Abyssal Plain

The Mariana Trench located in the western Pacific Ocean is the deepest trench on earth.

Model the Ocean Floor

Use an empty aquarium or other glass or clear plastic container and modeling clay or play dough to form a model of the ocean floor. The clay should be thick near the edge to represent the beach. Then it should slope down toward the center. In the center you can build mountains, valleys, and flat areas. Then put water into the aquarium. The surface of the water looks smooth even though the floor is not.

This underwater mountain sticks up above the surface of the water, forming an island.

Ocean Floor Worksheet

Label the parts of the ocean floor, then color the picture.

continental shelf

continental slope

abyssal plain

? What is the name of the part of the ocean floor that gently slopes away from the shore?

? What is the continental slope?

? What is the name of the flat part of the ocean floor?

Ocean Zones

The ocean is full of plants, algae, and animals. Most of these live in water where the sunlight reaches. Plants and algae need sunlight to grow. Many animals in the ocean eat the plants and algae, so they live where their food grows. This part of the ocean is called the

sunlit

zone. You will find sharks, coral, jellyfish, most whales, and most fish in this zone.

If you go deeper, you reach the twilight

zone. Only a little sunlight reaches this part of the ocean. There are no plants or algae in this zone. But some interesting animals live here. You might see a sperm whale, an octopus, and sponges.

Ocean Mural

Create an ocean mural for the wall. Tear off a length of white mural paper or butcher block paper. Use crayons to draw and color a scene of the ocean floor and the three zones. Include appropriate plants and animals in each zone. Use other resources to find as many different plants, algae, and creatures as you want to include. (You can work on this for several days and add to it as you learn more in the rest of the unit.) When the drawing is finished, use blue watercolors to paint over the whole scene, lighter at the top and getting darker toward the bottom. Display your ocean mural on a wall.

One of the most interesting animals found in the twilight zone is the lantern fish, which can actually glow in the dark.

Below the twilight zone is the _midnight_

zone. No light reaches the midnight zone. Only a few types of animals live this deep in the ocean. Some eels, anglerfish, and sea spiders have been found in this dark area. Most of these animals eat the bits of dead algae and animals that sink down to the ocean floor.

Sea spider

? Why do most sea creatures live in the sunlit zone?

? Name one animal you might find in the twilight zone.

? Why are there no plants in the midnight zone?

✏️ Scripture Trace

Praise the LORD from the earth, you great sea creatures and all deeps.   Psalm 148:7

Counting Sea Creatures Worksheet

Count each type of ocean animal. Write the number in the box. Then color the pictures.

angel fish		baby sea turtles		jellyfish	
octopi		sea horses		sharks	
shrimp		sponges		starfish	
whale					

Vents and Smokers

In certain areas on the ocean floor there are unusual features called _vents_. These vents are found in the midnight zone. Very hot water flows out of these openings in the ocean floor. Sometimes this water looks dark from the minerals in it. Since it looks like smoke, some vents are called black smokers. This water contains a mineral called _sulfur_.

There are special kinds of bacteria that lives around these vents. These bacteria can eat sulfur. Tube worms live in large colonies near these vents and eat the bacteria. Other animals have also been found living near the vents. Some of these include giant clams, white crabs, and lobsters. This part of the ocean shows God's creativity.

Tube Worm Dessert

Tube worms live near smoking vents on the ocean floor. Make a delicious model of a smoking vent. Scoop some ice cream into a bowl. Shape it like a small mountain. This is the vent. Top this with whipped cream to represent the smoke coming out of the vent. Drizzle chocolate syrup on the sides of the vent to represent the sulfur. Spread gummy "tube worms" around the vent. Enjoy eating your tube worm dessert.

A colony of tube worms living near a black smoker vent

? What is an ocean vent?

? What mineral is found in this hot water?

? How can animals live near vents when no plants or algae grow there?

Scripture Trace

If I take the wings of the morning and dwell in the uttermost parts of the sea, even there your hand shall lead me, and your right hand shall hold me.   Psalm 139:9–10

Ocean Vents Worksheet

Draw a line from each animal in the picture to its name. Look up the animals in a book, or have your teacher help you look up animals on the internet.

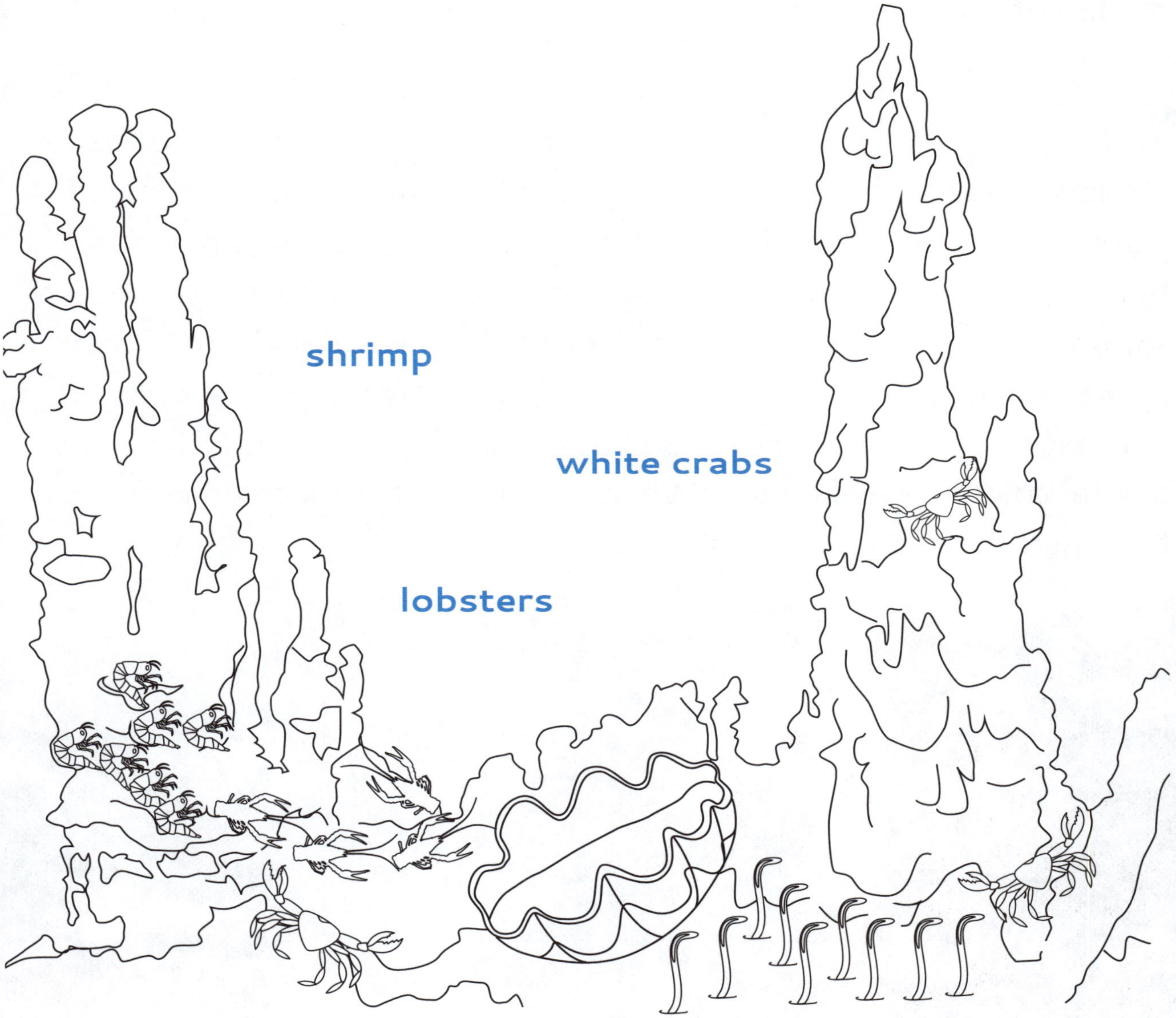

shrimp

white crabs

lobsters

tube worms

giant clam

Coral Reefs

One of the most beautiful areas in the ocean is a coral reef. A

coral reef is an underwater island. It is made by tiny creatures called corals. How can tiny animals build an island? Corals are animals that have soft bodies. These creatures make homes for themselves by growing hard cups around their soft bodies for protection. Millions and millions of these tiny animals build their homes together. This forms a coral colony. When one coral dies, another coral builds its home on top of the empty home. This is how a coral colony grows bigger. When hundreds of coral colonies grow together, they form a coral reef.

Corals live in warm, clear water near the equator. Many fish and plants like to live near coral reefs. Starfish, sea urchins, and sea anemones also live near coral reefs. A coral reef is one of the most fascinating areas to visit. People like to scuba dive around coral reefs to enjoy their beauty.

There are many different kinds of coral. Each type of coral builds a different shaped colony. Some corals build colonies that are shaped like human brains. These are called brain coral. Other coral colonies are shaped like snakes, fans, feathers, or even the antlers of a deer!

? What is a coral reef?

? How is a coral reef formed?

? Where are most coral reefs found?

Coral Reef Dot to Dot

Connect the dots to see a fish that lives in the coral reef. Then color the picture. The coral reef has many brightly colored animals. What shapes of coral colonies do you see in the picture?

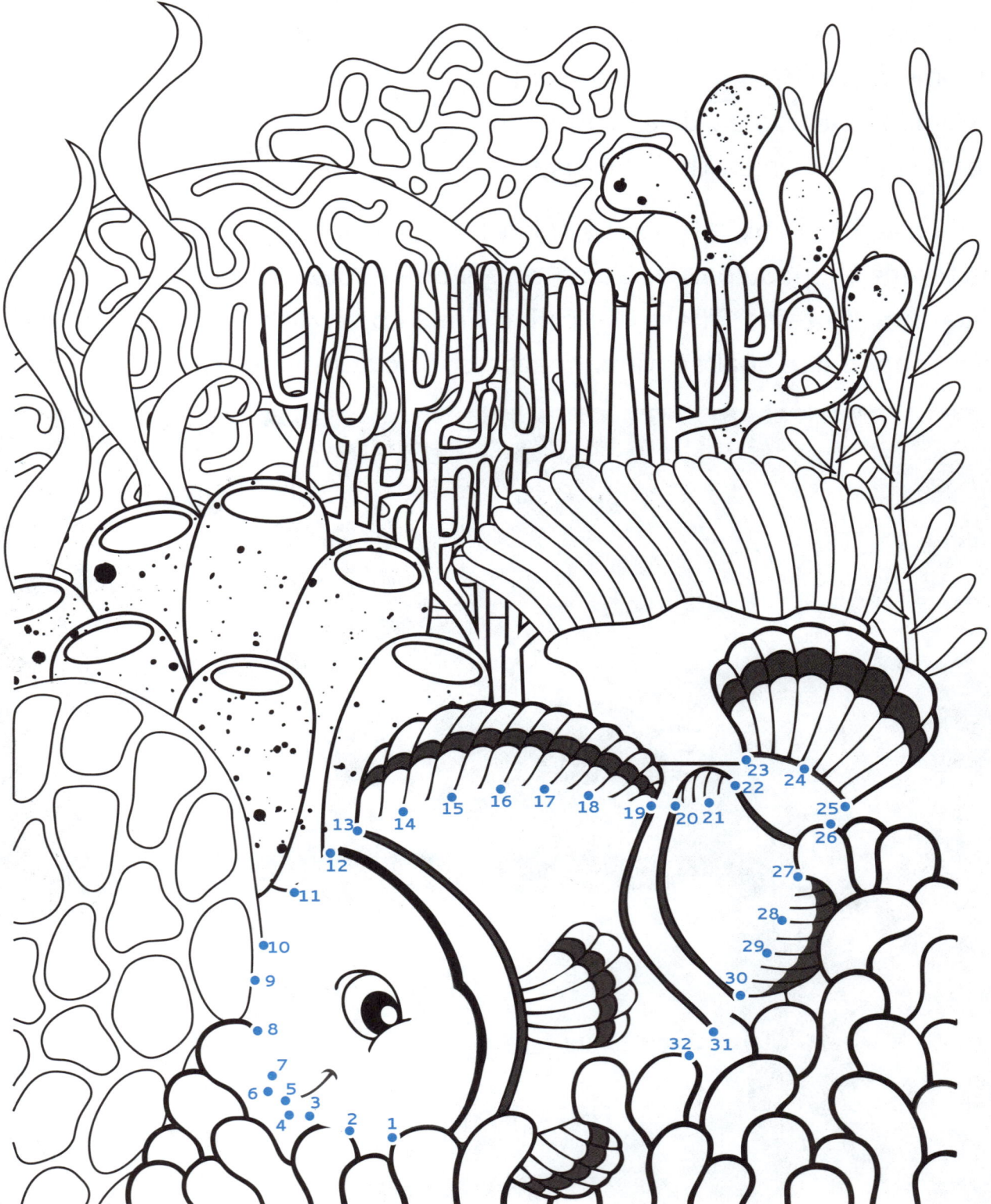

Conclusion

We live on a planet that was "very good" when God created it. But the earth was cursed when Adam and Eve sinned. This curse changed the earth. Later, God punished man for his wickedness by sending a great flood, which changed the earth even more. Even though the earth is changed, it is still a very special place.

The earth is mostly covered with water. This water makes life possible on Earth. Water from the oceans enters the air. This moisture is taken around the world by wind. God designed the weather that brings needed rain and sunshine. Look up in your own Bible and read the following Bible verses to see how God loves you and takes care of you through the weather. Then choose one verse to memorize and share with your family.

Psalm 19:1   **Psalm 24:1–2**   **Psalm 104:5–14**

? What is the most interesting thing you learned about weather?

? What is the most interesting thing you learned about the ocean?

✏️ **Scripture Trace**

May the glory of the LORD endure forever; may the LORD rejoice in his works.

Psalm 104:31

Unit Vocabulary Review

Match the vocabulary word to the correct picture:

Diving Suit

Air Tank

Submersible

Draw a line from the vocabulary word to the correct part of the seafloor:

Continental Shelf

Continental Slope

Abyssal Plain

Draw a line from the vocabulary word to the correct part of the ocean:

Midnight Zone

Sunlit Zone

Vent

Twilight Zone

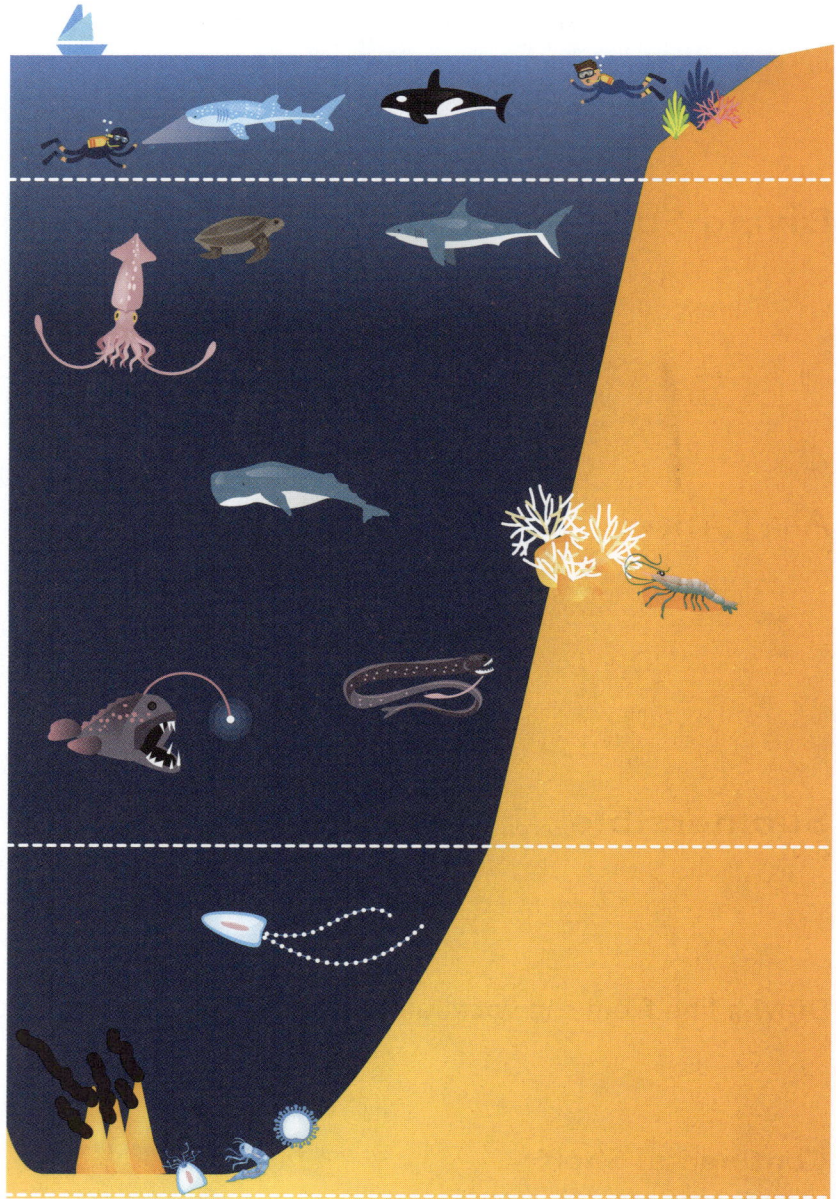

Fill in the blanks below.

1. S_____ is the mineral that is found in ocean vents.

2. A c_____r_____ is made by millions of tiny animals that build hard shells around themselves.

Universe
for Beginners

Space Models and Tools

Universe
for Beginners

Lessons 1–4

GOD'S
DESIGN®

Introduction to Astronomy

We are getting ready to learn all about space and the things that God has put there. This study is called

_____

astronomy . You are going to learn about the planets. You will learn about the sun, the moon, and the stars. And you will learn about our planet, Earth. All of these things were created by God.

? **What is astronomy?**

? **What questions do you have about space?**

✏️ **Scripture Trace**

For by him all things were created, in heaven and on earth.   Colossians 1:16

In the Beginning Worksheet

Read Genesis chapter 1 with your parent or teacher. This is the true story of creation. Look at the pictures of the different things that God made. Next to each picture write the day of creation on which that item was created.

sun

Created on

day _____

the earth
(dry land and seas)

Created on

day _____

moon

Created on

day _____

water animals

Created on

day _____

stars

Created on

day _____

land animals

Created on

day _____

light

Created on

day _____

the first man

Created on

day _____

The Earth Is Moving

Everything in the universe is moving. The earth moves in a circle around the sun. This is called

orbiting the sun. It takes one year for the earth to orbit the sun one time. The moon moves in a circle around the earth. It orbits the earth once each month.

The earth is also spinning. We call this spinning

motion _rotation_. The rotation of the earth is what makes the sun appear to rise and set. The earth spins all the way around one time each day. You don't feel the earth moving because you are moving with the earth.

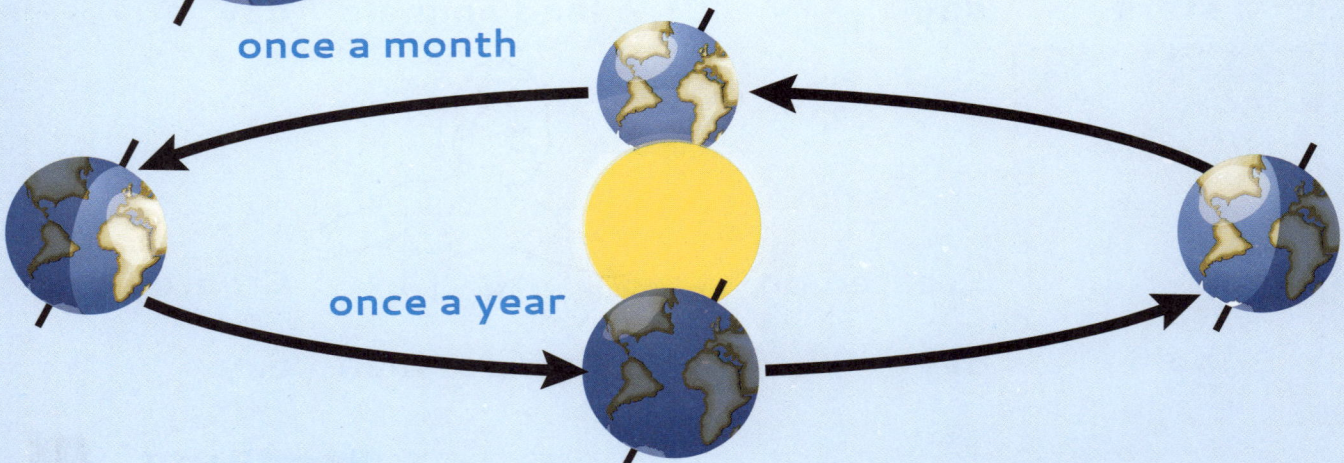

once a day

once a month

once a year

The force that keeps the planets moving around the sun is called

gravity. You cannot see gravity. But gravity is keeping you on earth. The earth's gravity pulls everything down. God created gravity to keep everything in its place.

What Moves Around the Sun?

On one piece of paper, draw a large yellow circle. This represents the sun. On a second piece of paper, draw a medium-sized blue circle. This represents the earth. On a third piece of paper, draw a small gray circle. This represents the moon. Have one person hold each piece of paper. The person holding the sun should stand in the middle of the room. The person holding the earth should slowly walk around the sun in a large circle. The person holding the moon should walk quickly in a small circle around the earth.

This exercise shows how the earth and the moon both move around the sun and how the moon also moves around the earth.

? How long does it take for the earth to go around the sun one time?

? How long does it take for the earth to spin around one time?

? What do we call the spinning motion of the earth?

? What force keeps all of the planets moving around the sun?

_ name _____

Day and Night Worksheet

As the earth rotates, a different part of the planet is facing the sun. It is daytime on the part of the earth that is facing the sun. The part facing away from the sun is dark. It is night there. Use a flashlight as the sun and a globe as the earth. Shine the flashlight on the globe. Notice how the part of the globe near the flashlight is lit up while the part away from the flashlight is darker.

Now look at the worksheet.

Write the word **Day** on the side of the earth that the sun is shining on.
Write the word **Night** on the side that is facing away from the sun. Shade the night side of the earth with a pencil.

_ *Universe Lesson 2* **139**

Why Do We Have Seasons?

As the earth moves around the sun over the year, we have different seasons. The earth is tilted in space. Look at a globe. You will see that the planet is not straight up and down. Now, place a large ball in the center of a table to represent the sun. Move the globe slowly around the ball without turning the stand. You will see that sometimes the top part of the globe is pointing toward the "sun" and sometimes it is pointing away from the sun. When the part of the earth where you live is tilted toward the sun, it is summer. The sun shines more directly there. This makes it warmer during the summer. During the winter, your part of the earth is tilted away from the sun. The sunshine comes to the earth at more of an angle. This makes it colder outside. When it is summer in the north part of the earth it is winter in the south part of the earth. The earth's tilt and

movement around the sun cause the _seasons_.

Spring   Summer

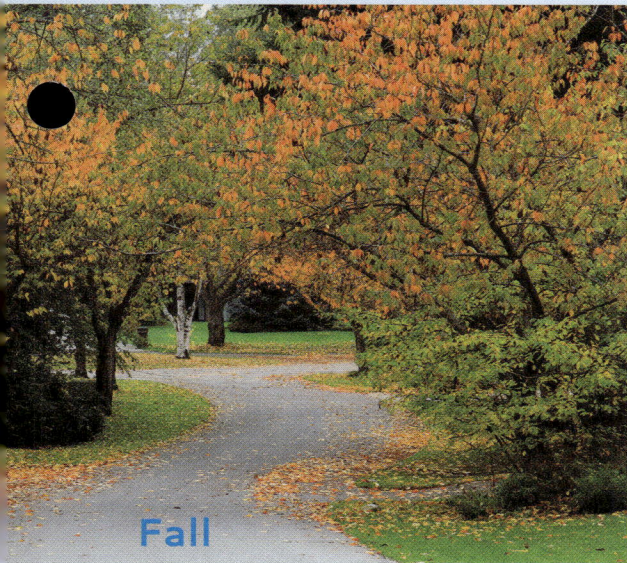

Fall

Winter

Musical Seasons

Listen to Vivaldi's *Four Seasons*. How does each movement of the music remind you of that season? Dance/move as if you were outside during that season. (For example, during winter, pretend you are walking in a snowstorm.)

? What are the four seasons?

? Why do we have different seasons?

Scripture Trace

And let them be for signs and for seasons, and for days and years.   Genesis 1:14b

Four Seasons Coloring Sheet

Because the earth is tilted, we get different amounts of sunshine at different times of the year. Most places have four different seasons each year. The spring is when new plants begin to grow. During the summer, it is warm, and you can do lots of fun things outside. In the fall the leaves on the trees change colors and fall off. In the winter it is often cold and sometimes snowy. Then spring comes again, and the cycle starts over. Each season has its own special things that you can do. Trace the name of each season, then color the pictures and think about what you like to do during each season.

Spring

Summer

Fall

Winter

Telescopes

Do you like to look at the stars? Many people do. The Bible tells us that "the heavens declare the glory of God" (Psalm 19:1). God created the beautiful night sky. We can see many stars just by looking up at night. But many people want to see more than just the lights in the sky.

Scientists are people who study the things God created. Some scientists study space. They have developed many special instruments to help them.

One of the most important instruments for looking at space is the

telescope. A telescope is like a special magnifying glass. It uses lenses and mirrors to make things look bigger. This allows us to see things in space much more clearly. With a telescope, you can see the surface of the moon. You can even see some of the planets in our solar system. Telescopes also help people see stars that are very far away.

Making Things Look Bigger

Use a magnifying glass to see how a lens can make things look bigger. A magnifying glass works the same way that a telescope works. A magnifying glass makes something that is small look bigger. A telescope makes something that is far away, like the moon or a star, look closer. If you have a telescope, use it to look at the moon. You can see the surface of the moon in great detail with a good telescope. If you do not have a telescope, you can use binoculars to look at the moon.

? Why do people like to use a telescope?

? How is a telescope like a magnifying glass?

What Can I See With a Telescope?

Color the things that a telescope would most likely help you see.

name _____

Unit Vocabulary Review

Two possible definitions are given for each vocabulary word below.
Draw a line through the wrong definition.

| Astronomy | The study of space |
| | The study of water |

| Orbiting | The earth spinning |
| | The earth moving around the sun |

| Rotation | The earth spinning |
| | The earth moving around the sun |

| Gravity | Force pulling down on things |
| | Force pushing things away |

| Seasons | Salt, pepper, cinnamon, nutmeg |
| | Summer, winter, spring, fall |

| Telescope | Instrument used to see things that are far away |
| | Instrument used to see things that are very small |

Outer Space

UNIT 2

Universe for Beginners

Lessons 5–10

GOD'S DESIGN®

Overview of the Universe

The universe is gigantic. No one really knows how big it is. There are billions of stars in the

universe.

Scientists believe that there may be too many stars in the universe to count. But one star is special to us. It is the sun. The sun is the closest star to Earth. It gives light and heat to the earth. There are eight planets that orbit the sun. The sun, the planets, and all the moons around these planets

make up our solar system.

✏️ Scripture Trace

Look toward heaven, and number the stars, if you are able to number them.   Genesis 15:5

Finding the Dippers

Sailors have used the stars to help them find their way on the sea. In order to use the stars, people have pointed out pictures that can be made by connecting certain groups of stars together. This is like a dot-to-dot puzzle. A group of stars that makes up a picture is called a

constellation. Look at the pictures showing the Big Dipper and the Little Dipper. These are two of the easiest groups of stars to spot in the night sky. Go outside after dark. Try to find a place with no streetlights or other lights. Look at the stars. Can you spot the Big Dipper and the Little Dipper?

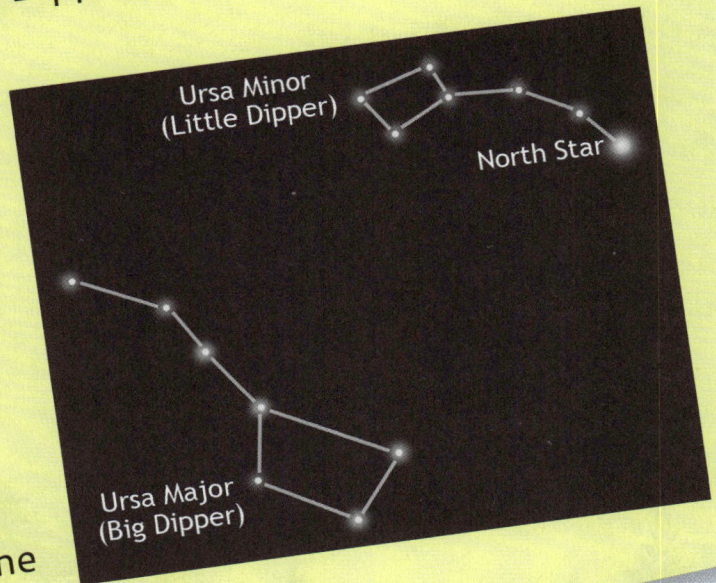

Ursa Minor (Little Dipper)

North Star

Ursa Major (Big Dipper)

? How big is the universe?

? Why is the sun a special star?

? What is the solar system?

Constellation Dot-to-Dot

Connect the dots to see Orion the Hunter.

1
2
3
4
5
6
7
8
9
10
11
12
13

Make Your Own Constellation

Choose an animal or object that you like. On dark blue or black construction paper, draw a simple outline of your animal or object with a white crayon. Place star stickers along the outline of your picture. Now give your constellation a special name.

Stars

Have you ever sung "Twinkle, twinkle little star, how I wonder what you are?" People often wonder what stars are. The **sun** is the closest star to Earth. Scientists have learned a lot about stars by studying the sun. They have learned that stars are not solid like the planet we live on. Instead, stars are made of super-hot gas. This is why they give off so much light and heat.

Our sun seems very big compared to all the other stars. But that is because it is much closer to us than any of the other stars. Using telescopes, scientists have discovered that some stars are much bigger than our sun. Other stars are smaller. Also, some stars are much brighter than our sun. Others are dimmer. They have also learned that our sun is just right to allow life to exist on Earth. It is just the right distance, brightness, and size for us. God knew what he was doing when he made all of the stars.

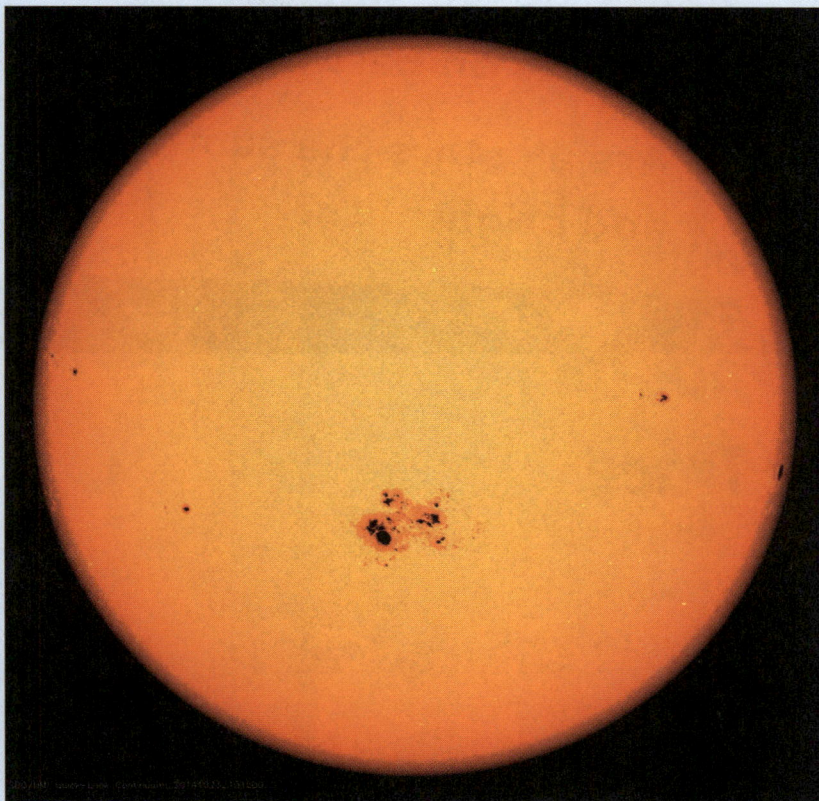

The sun

What Moves Around the Sun?

Research the names of real stars. You can use the internet (with your parent's permission) or an astronomy book. If you could name a star, what would you name it? What size would it be compared to the sun? How bright would it be compared to the sun?

? **What are stars made of?**

? **Why is our sun just right for life on Earth?**

? **Are all stars the same size and brightness?**

Scripture Trace

For star differs from star in glory.   1 Corinthians 15:41b

Star Worksheet

Smaller stars are usually cooler and give off a red light. Medium-sized stars, like our sun, usually have a yellow light. Larger stars are often hotter than other stars. They give off a blue light. Color the small stars **red**, the medium stars **yellow**, and the large stars **blue**.

Our Galaxy

We live on the planet earth. Earth orbits a star that we call the sun. All of the planets and moons that orbit the sun are part of our solar system. Our solar system is part of a very large group of stars. A large group of stars is called a _galaxy_. All of the stars in a galaxy spin around a central point. Our galaxy is called the _Milky Way_. The Milky Way is a spiral galaxy with five arms spinning around the center.

There are billions of other galaxies in the universe. Some galaxies are oval-shaped. Many are spiral-shaped like the Milky Way.

The giant oval-shaped galaxy ESO 325-G004.

Milky Way Picture

On a piece of black construction paper, use a white crayon to draw a pinwheel shape with five arms spiraling out from the center of the paper. Cover each arm and the center with glue and sprinkle sand on top. This is the shape formed by the stars in the Milky Way. You can see in the drawing that our sun is located about 2/3 of the way down one of the spiral arms.

Spiral galaxy Milky Way. The yellow dot is the location of our sun.

? What is a galaxy?

? What is the name of the galaxy we live in?

? What are two different shapes of galaxies?

The Sombrero galaxy is an oval-shaped galaxy with a flat disk tucked inside.

Milky Way Maze

Find your way through the Milky Way to reach the center of the galaxy.

Asteroids

Planets orbit the sun. But there are other things in space besides planets. An **asteroid** is a large rock that orbits the sun. An asteroid is much too small to be a planet.

Most of the asteroids in our solar system are in the space between Mars and Jupiter. This area is called the **asteroid belt**. There are many thousands of asteroids there. About 3,000 of these asteroids are big enough to receive a name. One of the asteroids is much larger than all the rest. It is round like a planet but still much smaller. It is called a dwarf planet. Its name is Ceres (SEER–eez). It is one of just five known dwarf planets in our solar system.

Ceres

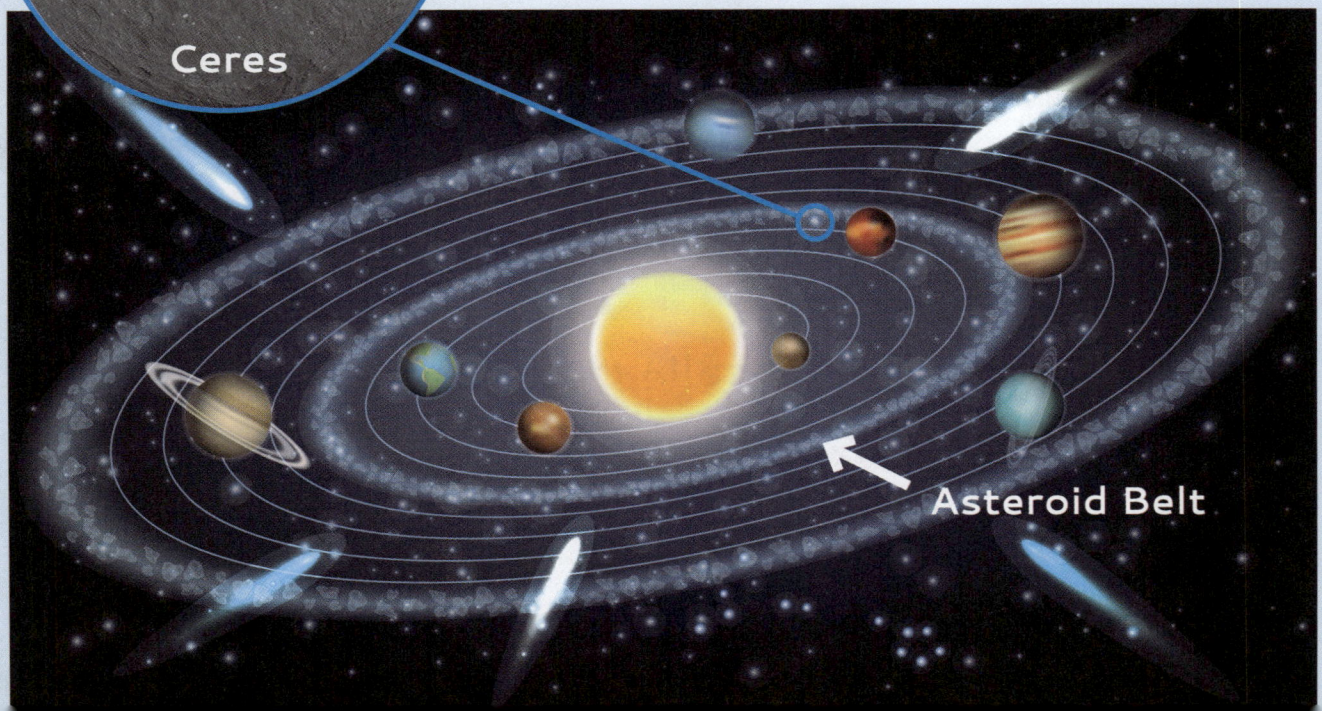

Asteroid Belt

Make an Asteroid Belt

Asteroids are giant rocks. They can have many different shapes. Some asteroids are smooth while others can be very rough. Asteroids are many different sizes. Some are much larger than others. Gather a large ball and four smaller balls. Then, go outside and collect some rocks. Gather many different shapes and sizes of rocks. The rocks should all be smaller than the balls you are using for this activity.

Place the large ball on the ground. This represents the sun. Place the 4 smaller balls going out from the sun. These represent Mercury, Venus, Earth, and Mars. Now use the rocks to make an asteroid belt. Make a ring that goes around all of the balls. These four planets are sometimes called the inner planets because they are inside the asteroid belt.

? What is an asteroid?

? Most asteroids are found between which two planets?

? What is this area called?

? What is the name of the largest asteroid?

Asteroid Coloring Sheet

Trace the word showing what asteroids are made of.
Then color the picture.

I am made of _rock_.

Adria Asteroid

Comets

Comets

are another kind of object that orbits the sun. Comets are different from asteroids. They are made mostly of ice. There are bits of rock and dust mixed in with the ice.

As a comet gets closer to the sun, it begins to melt. The ice turns into gas. Some of the gas and tiny pieces of dust are pushed away from the comet. This makes the comet look like a bright ball with a tail. The bright part of the comet is called the _head_. The pieces of dust that are flowing away from the sun form the _tail_.

Comet Model

You can make a model of a comet. Cut a small Styrofoam™ ball in half and glue it to a piece of tagboard. This is the head of the comet. Use glue on the tagboard to make a tail spreading out behind the comet. Cover the glue with glitter.

Tail

Head

? What is a comet made of?

? What happens when a comet gets close to the sun?

? What are the two parts of a comet?

Comet Coloring Sheet

Trace the word showing what comets are made of.
Then color the picture.

I am made mostly of ice.

Connor Comet

Meteors

Have you ever seen a shooting star? It looks like a bright light streaking across the sky. What you saw was probably a meteor.

A _meteor_ is a rock that has gotten too close to the earth. Gravity pulls it toward the earth. The rock passes through the atmosphere, which is the air surrounding the earth. As the rock goes through the earth's atmosphere it gets hot and burns up. This is why it is so bright.

Many meteors are the remains of comets. Comets melt a little every time they pass the sun. Eventually, comets break apart. This leaves bits of dust, rock, and ice floating in space. Sometimes the earth passes near the area where a comet broke up. Some of the pieces are pulled toward the earth. The ice melts. The dust and rocks burn up. And you might see a shooting star.

Geminids Meteor Shower

Once in a while a meteor does not get completely burned up. When this happens it hits the ground. Then we call it a meteorite. This crater was made by a meteorite. It is called the Barringer Crater. It is in Winslow, Arizona.

? What is a meteor?

? Why does a meteor become a shooting star?

Meteor Coloring Sheet

Trace the word showing what meteors are made of.
Then color the picture.

I am made of __rock__.

Mason Meteor

Unit Vocabulary Review

Fill in each blank with the correct vocabulary word from the following list. The first letter of each word is given for you.

asteroid, belt, comet, constellation, galaxy, head, meteor, Milky Way, solar, sun, tail, universe

1. The U_____ contains billions of stars.

2. Our S_____ system is made up of all of the planets and moons that orbit the sun.

3. A C_____ is a group of stars that forms a picture.

4. The S_____ is the closest star to the earth.

5. A g_____ is a large group of stars that all move together.

6. Our galaxy is called the M_____ W_____.

7. An a_____ is a large rock that orbits the sun.

8. Most asteroids are located in the asteroid b_____.

9. A c_____ is a ball of ice, dust, and rock that orbits the sun.

10. The bright part of a comet is the h_____.

11. The dust and gas that spread out from a comet form its t_____.

12. A m_____ is a rock that burns up in the earth's atmosphere.

Sun and Moon

Universe
for Beginners

GOD'S
DESIGN®

Lessons 11–18

Our Solar System

Our solar system includes the sun and everything that moves around it. There are planets and moons. There are asteroids and comets. There are also dwarf planets.

The sun is in the center of our solar system. Eight planets move around the sun. Four planets are small, and four planets are big. The four small planets are closest to the sun. Their names are Mercury, Venus, Earth, and Mars. Then comes the asteroid belt. The dwarf planet Ceres is in the asteroid belt. The four large planets are past the asteroid belt. Their names are Jupiter, Saturn, Uranus, and Neptune. There are more dwarf planets that are even further away from the sun than Neptune. Two of them are called Pluto and Eris.

SEE OPTIONAL ACTIVITY

Learning the Names of the Planets

You can learn the names and order of the planets (including Pluto and Eris the dwarf planets) by learning the following song — sung to the tune of "Twinkle, Twinkle Little Star."

Mercury, Venus, Earth, and Mars
These are planets that dwell with the stars.
Jupiter, Saturn, Uranus,
Neptune, Pluto, and Eris.
Mercury, Venus, Earth, and Mars,
These are planets that dwell with the stars.

Sun

Mercury

Venus

Earth

Mars

Saturn

Pluto

Asteroid Belt

Neptune

Uranus

Jupiter

Eris

? Name three things that are in our solar system.

? How are the four planets that are closest to the sun different from the four planets that are farthest away?

? Name two dwarf planets.

Solar System Coloring Sheet

Trace the name of each planet, the sun, and the asteroid belt, then color this picture of the solar system.

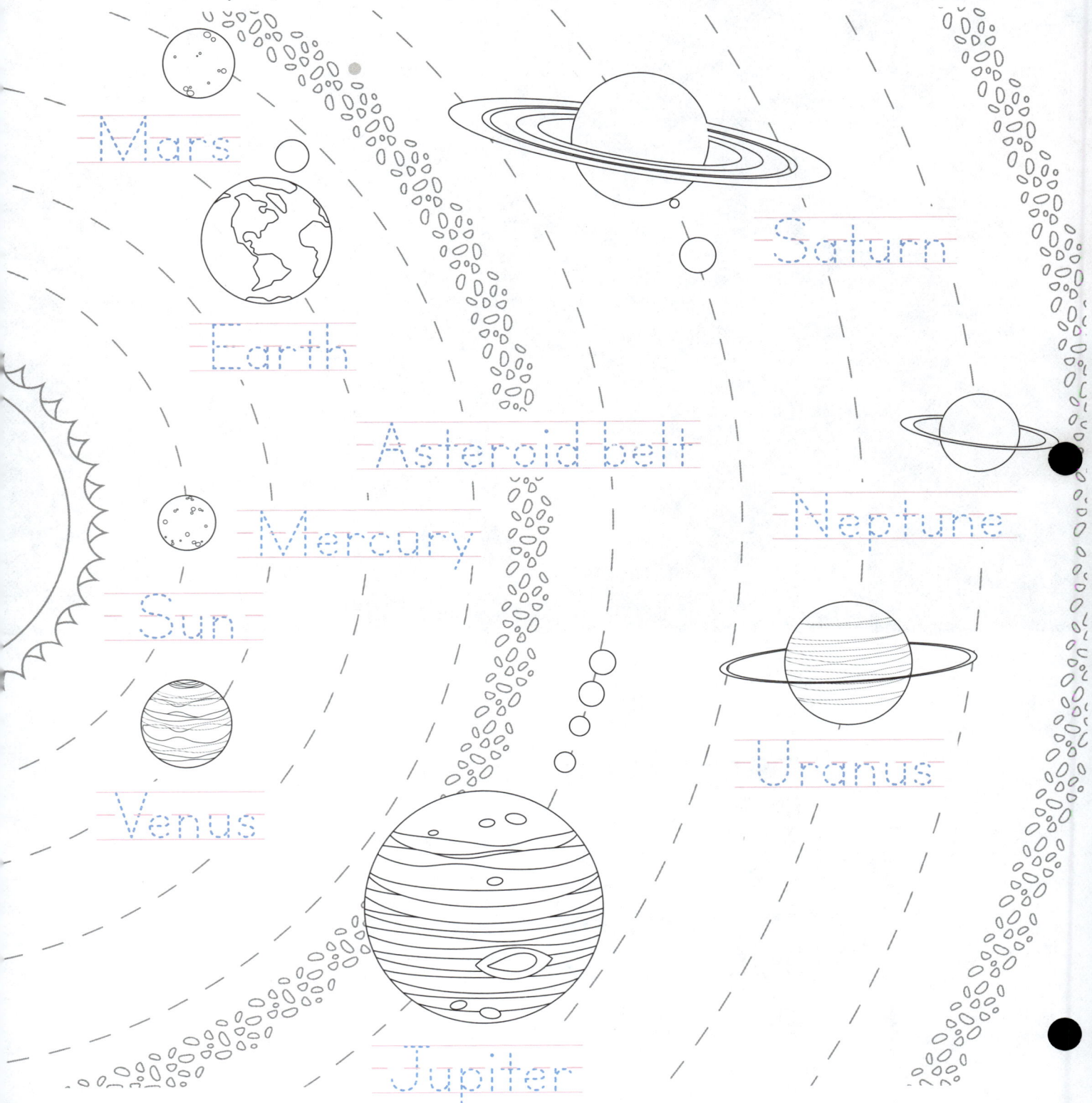

Mars

Earth

Asteroid belt

Saturn

Neptune

Mercury

Sun

Uranus

Venus

Jupiter

Our Sun

When you wake up in the morning, you usually see the bright sunshine. The sun is in the center of our solar system. Our sun is a medium-sized star. It has a somewhat yellow color. The sun is much larger than the earth. It is so big that one million Earths could fit inside the sun.

God designed the sun to give light and heat to the earth. Without the sun, no plants or animals could live on Earth. The sun is just the right distance away to provide enough heat to warm the earth and enough light for plants to grow.

Scripture Trace

The sun rises, and the sun goes down.   Ecclesiastes 1:5

The Colors of Sunlight

Sunlight looks mostly white with just a hint of yellow. But sunlight is actually all different colors of light mixed together. That is why we can see so many different colors in a rainbow. As sunlight passes through the water drops in the sky, the different colors of light are split up. This forms a rainbow. You can make your own rainbow by doing the following:

1. Fill a pie pan with water and place it on a level surface near a window.

2. Place a small mirror in the water so that it reflects the sunlight onto a wall after the sunlight has passed through the water. You should see the colors of the rainbow on the wall.

? Which is larger, the sun or the earth?

? How big is our sun compared to other stars?

? Why do we need the sun?

Colors of Light

When light goes through a prism, you can see the different colors within the light, just like you do when the sun shines through the drops of rain and forms a rainbow. See if you can color the beams of light as they leave the prism. To remember what order the colors appear in a rainbow or a prism, you can use the letters in the name Roy G. Biv.

R O Y G B I V

blue   yellow

violet   red   green

orange   indigo

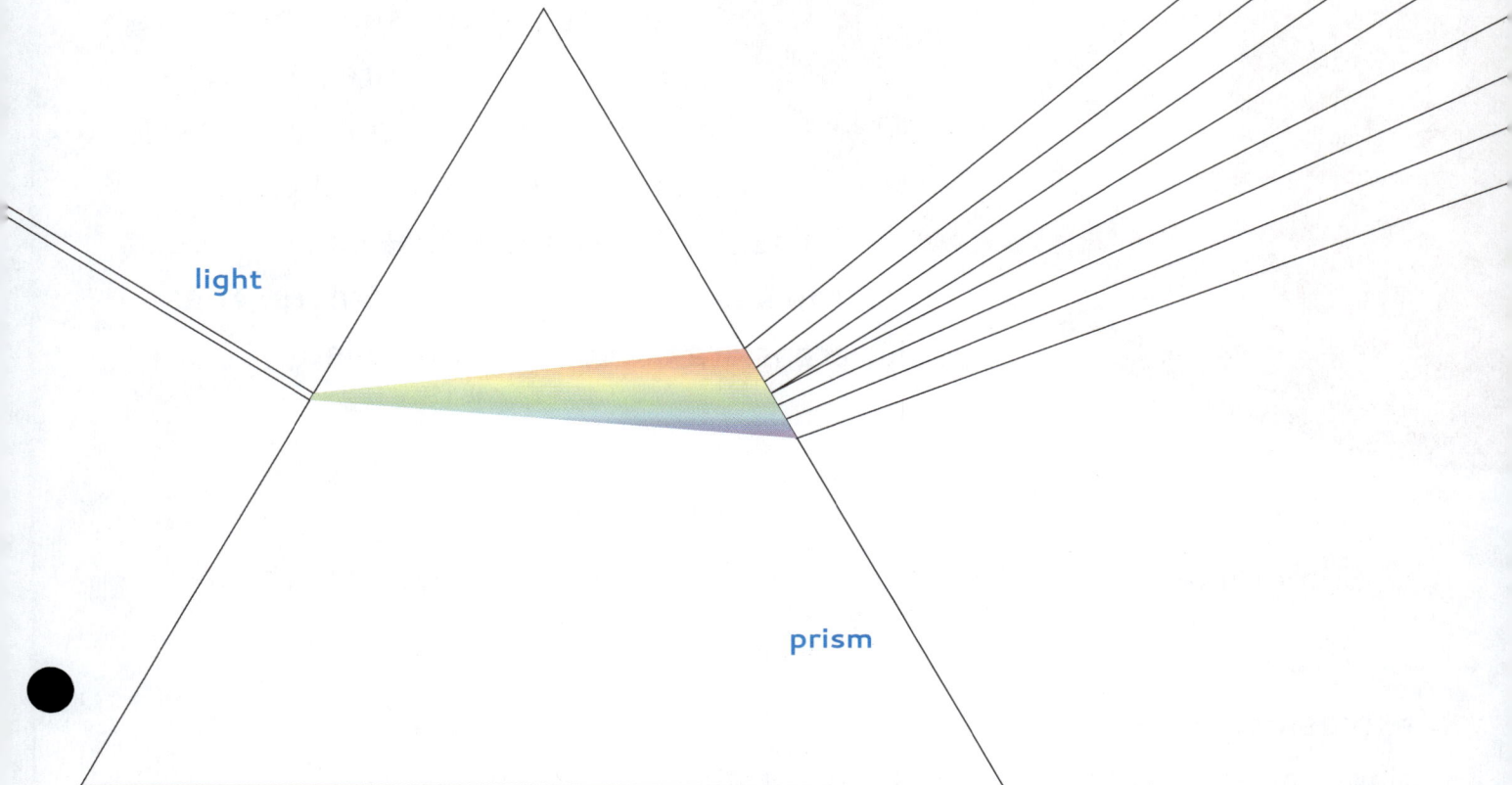

light

prism

The Surface of the Sun

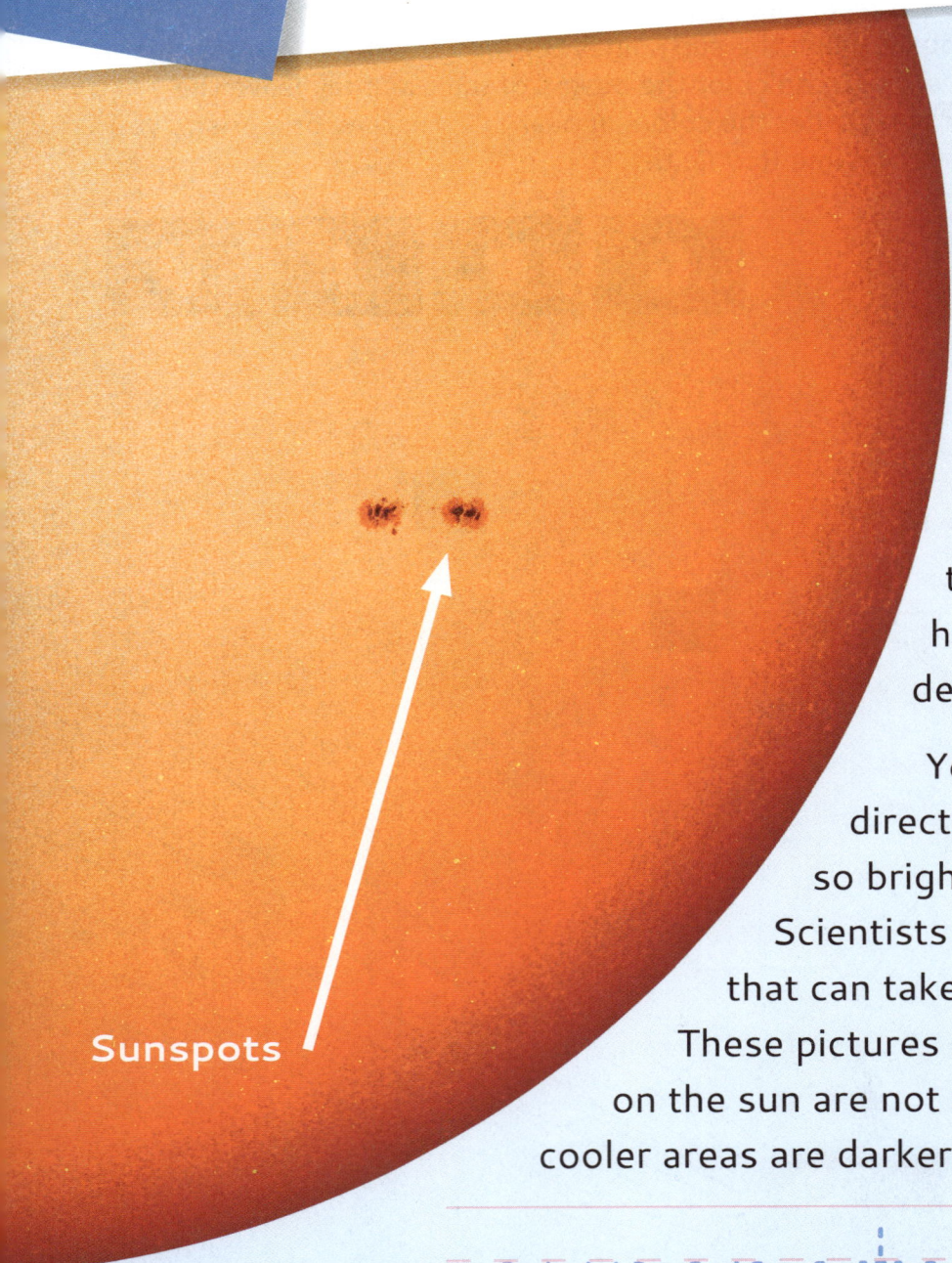

Sunspots

The sun is a star. It provides heat and light to the earth. The sun is made from mostly hydrogen and helium, which are two gases. The surface of the sun is extremely hot. It is over 10,000 degrees!

You should never look directly at the sun. The sun is so bright it can hurt your eyes. Scientists have special cameras that can take pictures of the sun. These pictures show that some places on the sun are not as hot as others. These cooler areas are darker and are called

sunspots.

Sometimes there are explosions on the surface of the sun. These explosions are called solar flares. A solar flare can send material thousands of miles into space.

Tracing Your Shadow

You can have fun tracking the movement of the sun through the sky as the earth rotates.

1. On a sunny day go outside early in the morning and make an X on the sidewalk with a piece of sidewalk chalk.

2. Stand on the X and have someone trace your shadow on the sidewalk.

3. Go outside every 2–3 hours throughout the day and repeat this. Make sure you stand in the same place each time.

You will see your shadow move as the sun moves across the sky.

? Why should you never look directly at the sun?

? What is a sunspot?

? What is a solar flare?

Solar flare on the surface of the sun.

Sammy Sun Coloring Sheet

Trace the words and color the picture of Sammy Sun.

sunspot

solar flare

Sammy Sun

Solar Eclipse

The earth is moving around the sun. The moon is moving around the earth. Sometimes the moon comes directly between the earth and the sun. If the moon is in just the right place, it can block the light from the sun. When this happens, we call it

a __solar eclipse__.

You have to be in just the right place to watch a solar eclipse. The darkness from the eclipse is only about 150 miles (240 km) wide. You should not look directly at the sun even during an eclipse. Although the moon blocks most of the light from the sun, it does not block all of the harmful light. You can still hurt your eyes. You can safely watch an eclipse if you wear specially designed eclipse glasses like the ones this kid is wearing.

moon

sun

During an eclipse the moon begins to cover the sun. It slowly moves across the sun until the sun is totally blocked. The light from the sun gets blocked for a few minutes. Then the moon slowly moves away from the sun, and it becomes bright again.

Moon Shadow

Use a globe or large ball for the earth, a small ball for the moon, and flashlight for the sun. Go into a dark room. Shine the flashlight onto the globe or ball. Slowly move the small ball into the beam of light, passing between the globe and the flashlight. Notice the shadow of the small ball on the globe as you move the ball. This is what happens during a solar eclipse.

? What causes a solar eclipse?

? During an eclipse is the whole earth dark?

Make Your Own Solar Eclipse Worksheet

You can make your own solar eclipse. Color the moon gray and the sun yellow. Cut out the sun and the moon. Slowly slide the moon in front of the sun. Notice how it blocks out some of the light as it covers up the sun. Move it until the moon completely covers the center of the sun. This is a total eclipse. Keep moving the moon slowly until it is no longer in front of the sun.

Blank for cutting

Solar Energy

The heat and light from the sun help us in many ways. Its heat warms the earth. It gives light for people and animals during the day. The light helps plants to grow and make food. Heat and light are two kinds of energy. Energy from the sun is called

solar energy.

People have learned how to use the heat and light from the sun to make useful energy. You may have seen

solar panels on the roof of a house. Solar panels can store heat from the sun. This heat is used to heat water. Special

solar cells

can turn sunlight into electricity. You may have used a calculator that uses the light from the sun instead of batteries. People are trying many different ways to use the sun's energy.

? **What is solar energy?**

? **What types of energy come from the sun?**

? **What are two things that people have invented to help them use solar energy?**

Solar Energy Worksheet

○ Grass uses sunlight to grow. Circle everything else in the picture that uses sunlight.

Our Moon

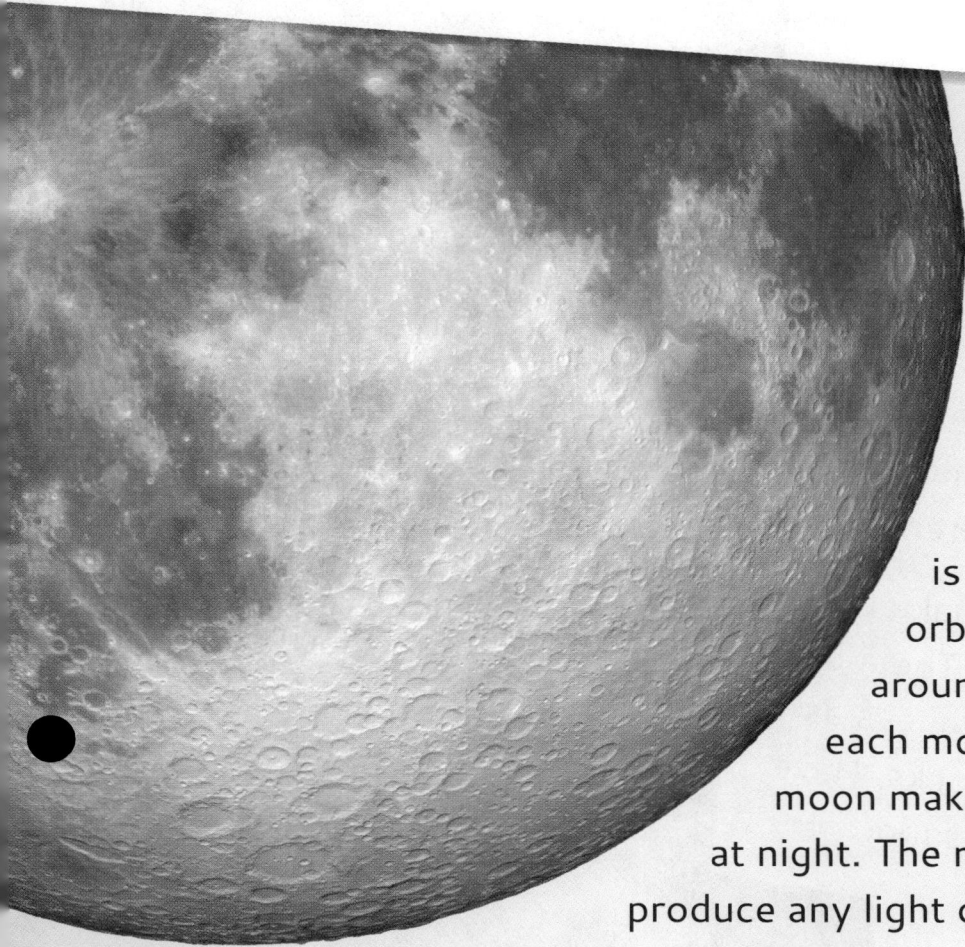

The __moon__ is a large round rock that orbits the earth. It goes around the earth about once each month. The light from the moon makes it easier to see things at night. The moon does not actually produce any light of its own. The moon __reflects__ light from the sun, which means the sun's light bounces off of the moon and comes to the earth. This allows the earth to have light at night as well as during the day.

The moon is much smaller than the earth and has less gravity, which means the moon does not pull down as hard on things as the earth does. So, when astronauts visited the moon, they were able to leap long distances. The surface of the moon is very different from the surface of the earth. There is no air to breathe. There is very little water so there are no living plants or animals. There are just rocks, hills, valleys, and craters on the moon. God did not create the moon for us to live on. But He did create it to give us light during the night.

Moon Sand

You can make fun moon sand using the following recipe.

4 cups of sandbox sand

2 cups of corn starch

1–3 cups of water — enough to make a smooth consistency

Use your sand to make the surface of the moon. The moon has lots of hills and valleys as well as craters and rocks. Have fun with your moon sand.

? How long does it take for the moon to go around the earth one time?

? Where does the light from the moon come from?

? Why are there no plants or animals living on the moon?

What Do I Weigh on the Moon?

How much would you guess that you weigh on the moon? Did you know that because gravity is different on the moon, your weight would be different? Remember, the moon's gravity is less than the earth's gravity. If you went to the moon, do you think you would weigh more or less? Circle your answer.

MORE   or   LESS

Now look at the chart below. Find the number closest to your weight. Look at the number next to it. That's about how much you would weigh on the moon!

Write your moon weight here: _____.
Are you surprised by your moon weight?

My weight on Earth	My weight on the moon
30 pounds	5.0 pounds
35 pounds	5.8 pounds
40 pounds	6.6 pounds
45 pounds	7.4 pounds
50 pounds	8.3 pounds
55 pounds	9.1 pounds
60 pounds	10.0 pounds

Phases of the Moon

The moon is always moving around the earth. Each day it is in a slightly different place compared to the sun. So, the light from the sun reflects off of it in a different way each day. Sometimes light shines

from the whole moon. This is called a _full moon_. Most of the time light only shines from part of the moon. Sometimes no light shines from the moon at all. When no light comes from the

moon, it is called a _new moon_. The next night a little light will reflect off the moon. The amount of light will grow each night until there is a full moon. After the full moon, the moon reflects less light each night until there is another new moon. The time between new moons is about one month.

Full Moon

✏️ Scripture Trace

Blow the trumpet at the new moon, at the full moon, Psalm 81:3

Phases of the Moon Cookies

You can have fun showing the phases of the moon with a few sandwich cookies. Remove the top cookie from 8 sandwich cookies revealing the frosting. This white frosting looks like a full moon. On two cookies remove about ¼ of the frosting. On two of the cookies remove about ½ of the frosting. On two of the cookies remove ¾ of the frosting. And on one cookie remove all of the frosting. This cookie represents the new moon.

Place the cookies on a paper plate in a circle showing the phases of the moon. Use the picture in this lesson as a guide.

? Why does the light shining on the moon change from day to day?

? What is a full moon?

? What is a new moon?

Where Did the Moon Come From?

Where did the moon come from? The Bible answers this question. According to Genesis 1:1–19, God created the earth on the first day of creation. Then he created the sun, moon, and stars on the fourth day of creation. He spoke, and they came into being from nothing. Aren't you glad God made the earth, sun, moon, and stars?

Some people do not believe in God. They try to explain where the moon came from in other ways. One idea says the moon first orbited the sun. Then something knocked it out of its orbit. The earth's gravity then pulled the moon into orbit around the earth. This idea is very unlikely. Trusting God's Word gives us a better answer for where the moon came from. God made it for His glory!

? Where did the moon come from?

? On which day of creation did God make the moon?

Scripture Trace

And God made . . . the lesser light to rule the night.   Genesis 1:16

What Is True About the Moon?

◯ Circle the statements that are true about the moon.

✗ Mark out the statements that are not true.

God created the moon.

The moon is made of cheese.

The moon is a planet.

The moon makes its own light.

The moon reflects the sun's light.

The moon revolves around the earth.

The moon is made of rock.

The moon is smaller than the earth.

Unit Vocabulary Review

Match the vocabulary word to its definition.

Sunspots   ○   Cool areas on
                          the sun's surface

Solar Flare   ○   Moon covers
                          the sun

Solar Eclipse   ○   Explosion on the
                          sun's surface

Solar Panel   ○   Creates electricity
                          from sunlight

Solar Cells   ○   Heats water
                          using sunlight

Moon   ○   How sunlight bounces
                          off the moon

Reflect   ○   When no sunlight
                          reflects off of the moon

Full Moon   ○   Large rock that
                          orbits the earth

New Moon   ○   When the whole surface
                          of the moon is lit up

Planets

Universe
for Beginners

GOD'S
DESIGN®

Lessons 19–27

Mercury

The closest planet to the sun is called

Mercury. Mercury is much smaller than Earth. It is the smallest planet in the solar system.

Mercury has no atmosphere. That means it has no air around it. Because Mercury is so close to the sun, the side facing the sun gets very, very hot. And because it has no atmosphere, the side facing away from the sun gets very, very cold. You wouldn't want to visit Mercury.

Atmosphere Is Important

The temperature on Mercury is either very hot or very cold but never comfortable. One reason for this is that Mercury does not have an atmosphere like Earth does. Let's see how an atmosphere helps to make temperatures more comfortable. (*Note: Adult supervision required.*)

Pretend that your arms are Mercury and Earth. Wrap a towel around the arm that is Earth. The towel acts like Earth's atmosphere. Now, the teacher can put the hair dryer on low, and then gently blow warm air on your arm. (*Be very careful not to let the blow dryer get too close or touch the student's arm.*) Think about how warm each arm feels. Earth's atmosphere helps to protect us from the sun's heat just like the towel protects your arm from the hot air.

Now place a piece of ice on each arm for a few seconds. How does each arm feel? Again, the atmosphere (the towel) protects Earth from the cold temperatures of space. You would not be able to live on a planet like Mercury that does not have an atmosphere.

? Why would it be a bad idea to visit Mercury?

? Is Mercury closer to the sun or farther away from the sun than earth is?

Mercury Facts Sheet

Trace the name and color the planet. Review the facts you have learned about this planet and fill in the words below. The first letter of the answer has been given to you.

Mercury

Mercury is the C_____ planet to the sun.

Mercury is the S_____ planet.

Mercury has no G_____.

Venus

Venus

is the second planet
from the sun. It is only a
little smaller than Earth. Venus has
an atmosphere, but you could not live on this
planet. The air around Venus is made of carbon dioxide. People need
air that has oxygen in it. Also, Venus is surrounded by thick, poisonous
clouds. You could not breathe the air. The atmosphere on Venus is
very thick and traps heat from the sun. Venus is the hottest planet in
the solar system. You would burn up if you went there.

The clouds that surround Venus reflect the light of the sun. This
causes Venus to look like a very bright star in the night sky. Venus is
often one of the first things you can see in the sky at night. It is also
one of the last things to disappear when the sun begins to rise. So,
Venus is sometimes called the Evening Star or the Morning Star.

Can You Spot Venus?

If you want to find Venus in the night sky, you can look toward the west after sunset or the east before sunrise each day. Venus should be the brightest thing in the sky other than the moon. It will look like a very bright star. Venus is usually visible for about three hours after sunset or about three hours before sunrise.

? **Why is Venus so hot?**

? **Why can't people live on Venus?**

? **Why is Venus called the Morning Star?**

Venus Facts Sheet

Trace the name and color the planet. Review the facts you have learned about this planet and fill in the words below. The first letter of the answer has been given to you.

Venus

Venus is just a little smaller than E_____.

Venus is the S_____ planet from the sun.

Venus is the h_____ planet in the solar system.

The atmosphere on Venus is P_____ to humans.

Earth

Earth is the third planet from the sun. It revolves around the sun once each year. It spins around once each day. The earth has one moon. That moon moves around the planet about once each month. The moon gives us light at night.

Earth is the planet that God made especially for people. God created everything just right so that plants, animals, and people could live here. It is just the right distance from the sun to keep us warm without making us too hot.

It has lots of water. Plants, animals, and people need water to live. No other planet in our solar system has much water. God also designed Earth to produce clouds. The clouds move water from one place to another. When this water falls as rain, it helps plants grow in nearly every part of the world.

Earth also has air. The air has just the right amount of oxygen for people to breathe. God designed plants to produce oxygen when they grow. Animals and people breathe in the oxygen. They breathe out a gas called carbon dioxide, which the plants then use as they grow. What an amazing recycling system! Earth is a very exciting planet because God made it just for us.

? List three ways that Earth is just right for people.

? Which planets are closer to the sun than Earth?

Just Right Game

You will need a six-sided die to play this game. Place this game board in front of you. Roll the die. Find the section number that matches the number on the die. Tell how the item in that section shows that God designed Earth just right for people to live on. Pass the die to the next person. Play continues until someone has had a chance to talk about all sections on the board.

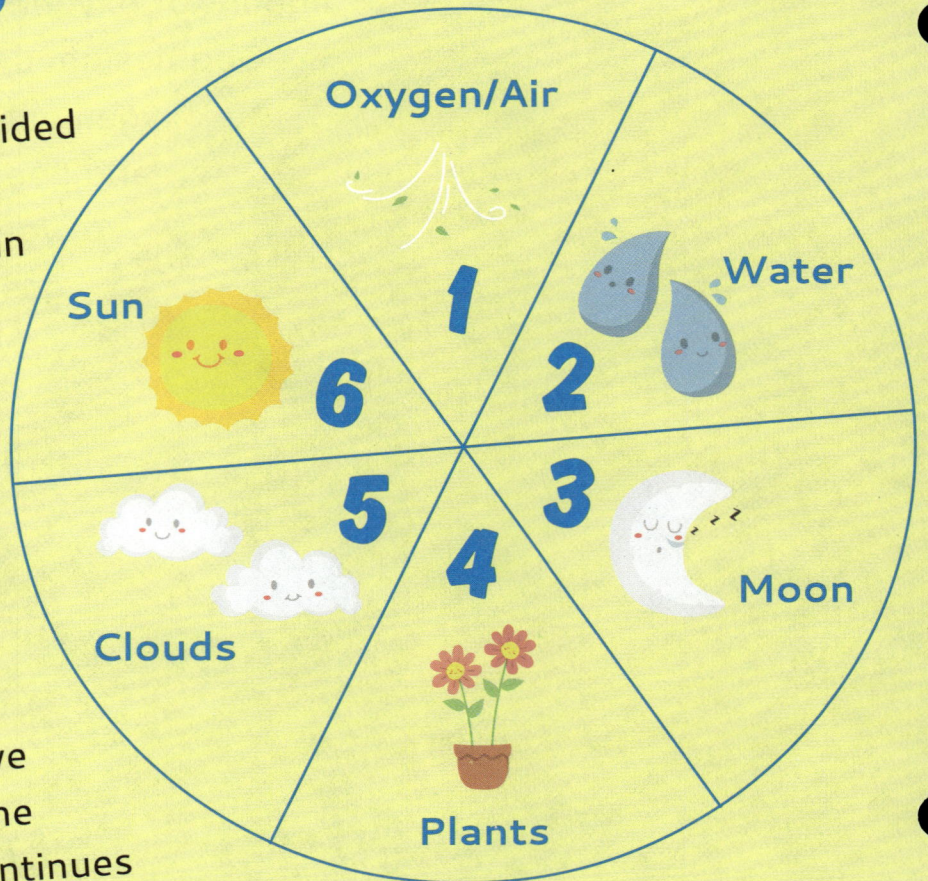

Oxygen/Air

Water

Sun

1

2

6

3

5

4

Moon

Clouds

Plants

Scripture Trace

And my God will supply every need of yours according to his riches in glory in Christ Jesus. Philippians 4:19

Earth Facts Sheet

Trace the name and color the planet. Review the facts you have learned about this planet and fill in the words below. The first letter of the answer has been given to you.

Earth

Earth is the t_____ planet from the sun.

Earth is designed just right for P_____ to live there.

Earth has a lot of W_____.

Earth has air with O_____.

Mars

Mars

is the fourth planet from the sun. Mars is about half as big as Earth. It is just a little bigger than Mercury. Mars is sometimes called the red planet. The soil has a large amount of rust in it. This gives the planet a reddish tint. Mars has a small amount of air around the planet. But the air is mostly carbon dioxide. It does not have the oxygen that you need to breathe.

Mars has a north pole and a south pole. Both of these poles have ice on them all the time. The ice on Mars is a little different from the ice on Earth. Some of the ice on Mars is made from water just like on Earth. But much of the ice is made of carbon dioxide. Frozen carbon dioxide is called dry ice.

Valles Marineris

We have learned many things about the surface of Mars. Mars has sand dunes, craters, glaciers, and volcanoes. It has mountains and canyons. One of these canyons is much larger than the Grand Canyon on earth. In fact, the Valles Marineris is the largest canyon in the entire solar system.

? What is the name of the fourth planet from the sun?

? Why is Mars sometimes called the red planet?

? Why would you not be able to breathe on Mars?

Take a Trip to Mars

Many people today believe it will someday be possible to send people to Mars. You can take a pretend trip to Mars. Circle all the things below that you would need to take with you on a trip to Mars. If you like, you can collect similar items from around your house to use as you pretend to go to Mars.

Now that you have all of your supplies you can start your trip. Choose a location in your house to be your spaceship. Perhaps you can build a spaceship by draping a blanket over some chairs. Get into your ship and fly to Mars. Once you get there, you can pretend to set up your camp. Then you can do some of the following things:

☐ Jump really high — you will weigh about 1/3 as much on Mars as you do on Earth.

☐ See lots of rocks — there are not any plants or animals on Mars, only dirt, rocks, and ice.

☐ Look for very large canyons — Mars has several large canyons. One canyon is much bigger than the Grand Canyon.

☐ Look for Earth in the night sky — it would look like a bright star.

Have fun exploring the Red Planet.

Mars Facts Sheet

Trace the name and color the planet. Review the facts you have learned about this planet and fill in the words below. The first letter of the answer has been given to you.

Mars

Mars is the f_____ planet from the sun.

Mars is a little bigger than M_____.

Mars is called the R_____ Planet.

Jupiter

The fifth planet out from the sun is

Jupiter.

Jupiter is the biggest planet in our solar system. It is much, much bigger than Earth. All of the planets that you have learned about so far are made of rock. But Jupiter and the other planets that are farther from the sun are different. They are giant balls of gas. There is no solid land on Jupiter.

One of the most famous features of Jupiter is its Great Red Spot. The Great Red Spot is believed to be a giant storm that has been raging for hundreds of years. The Great Red Spot is so big that two or possibly three planets the size of Earth would fit within it. It can be seen from Earth with telescopes and was first described in 1665.

There are at least 70 moons that orbit Jupiter.

Earth (left) compared in size to Jupiter (right)

How Big is Jupiter?

Use two cereal bowls put together to represent Jupiter. Imagine Jupiter is split into halves. Put Jupiter's two halves (the bowls) on the table. Use marbles or round cereal to represent Earth. Count how many "earths" can fit inside "Jupiter." After you fill one bowl, you can double that number. If you actually had round items the right size, you could fit 1,300 Earths inside of Jupiter.

? How is Jupiter different from Earth?

? What is the Great Red Spot?

Jupiter Facts Sheet

Trace the name and color the planet. Review the facts you have learned about this planet and fill in the words below. The first letter of the answer has been given to you.

Circle the Great Red Spot.

Jupiter

Jupiter is the _____ planet in the solar system.

Jupiter is the f_____ planet from the sun.

Jupiter is made of g_____ .

Jupiter has over 70 m_____ .

The Great Red Spot is a giant s_____ .

Saturn

Saturn is the sixth planet from the sun. It is the second largest planet. It is also a gas planet like Jupiter. The most famous feature of Saturn is its rings. There are thousands of small rings around Saturn. These rings are made from bits of ice, dust, and rocks.

Saturn also has at least 80 moons and maybe more! One of these moons is large. Six are medium-sized. The rest are small. The largest moon is called Titan. Titan is bigger than the planet Mercury.

? How many moons does Saturn have?

? What are Saturn's rings made of?

Saturn Facts Sheet

Trace the name and color the planet. Review the facts you have learned about this planet and fill in the words below. The first letter of the answer has been given to you.

Saturn

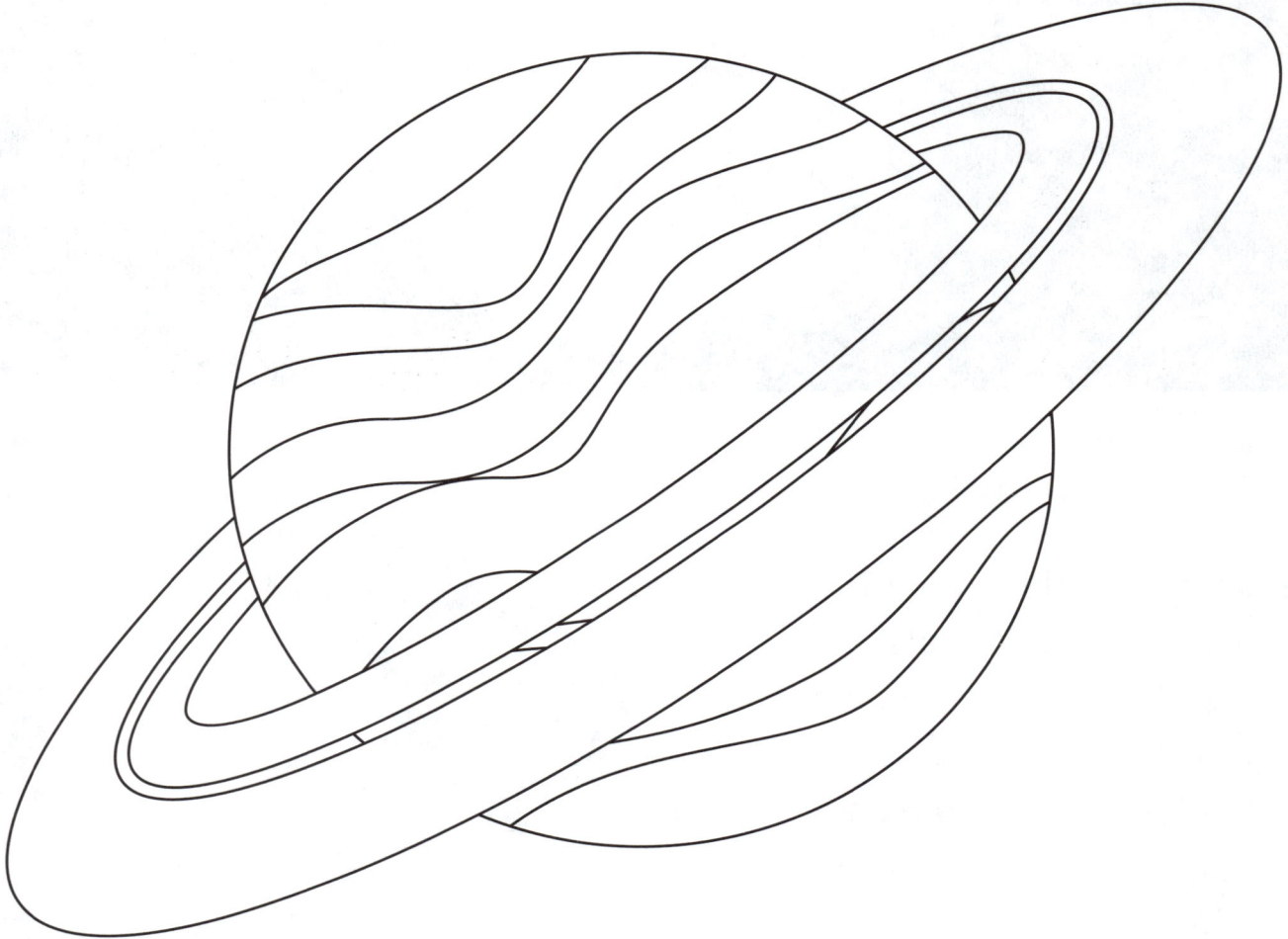

Saturn is the S_____ planet from the sun.

Saturn is a g_____ planet.

Saturn has thousands of r_____.

Titan is the largest m_____ that orbits Saturn.

Uranus

The seventh planet from the sun is __Uranus__. Uranus is a gas planet like Jupiter and Saturn. It is smaller than Saturn. But it is much larger than Earth. Uranus is a pale blue color.

The way that Uranus moves is different from other planets. Most planets spin like a top. If you put a stick through the center of most planets, the stick would go through the top and bottom. But Uranus rolls around the sun on its side. A stick through its center would go from side to side.

Uranus also has rings. However, there are only a few rings around Uranus. Uranus also has more than 20 moons. Most of these moons are small.

? How does Uranus move in a way that is different from the other planets?

? Is Uranus a gas planet or a rock planet?

How Does Uranus Move?

Push a popsicle stick or wooden skewer through the top to the center of an apple. Spin the apple slowly while holding the stick almost up and down. This is how Earth moves. Shine a flashlight on the spinning apple. Notice how one side of the apple is lit up and the other side is dark. As Earth spins the sun lights up part of it. It is daytime on that side of the planet. Then as Earth spins away from the sun that side becomes dark and it is night.

Now lay the apple down so that the stick goes side to side (see illustration of planet Uranus spinning below). Shine the flashlight at one end of the stick. Again, slowly spin the apple. Notice that the same side of the apple is always facing the light.

If you could live near one of the poles (the point where the stick comes out of the apple) you would have 42 years of sunlight. Then you would have 42 years of darkness. It is a good thing that Earth does not spin this way.

Now cut the apple in half to remove the stick and enjoy a delicious space treat.

Uranus Facts Sheet

Trace the name and color the planet. Review the facts you have learned about this planet and fill in the words below. The first letter of the answer has been given to you.

Uranus

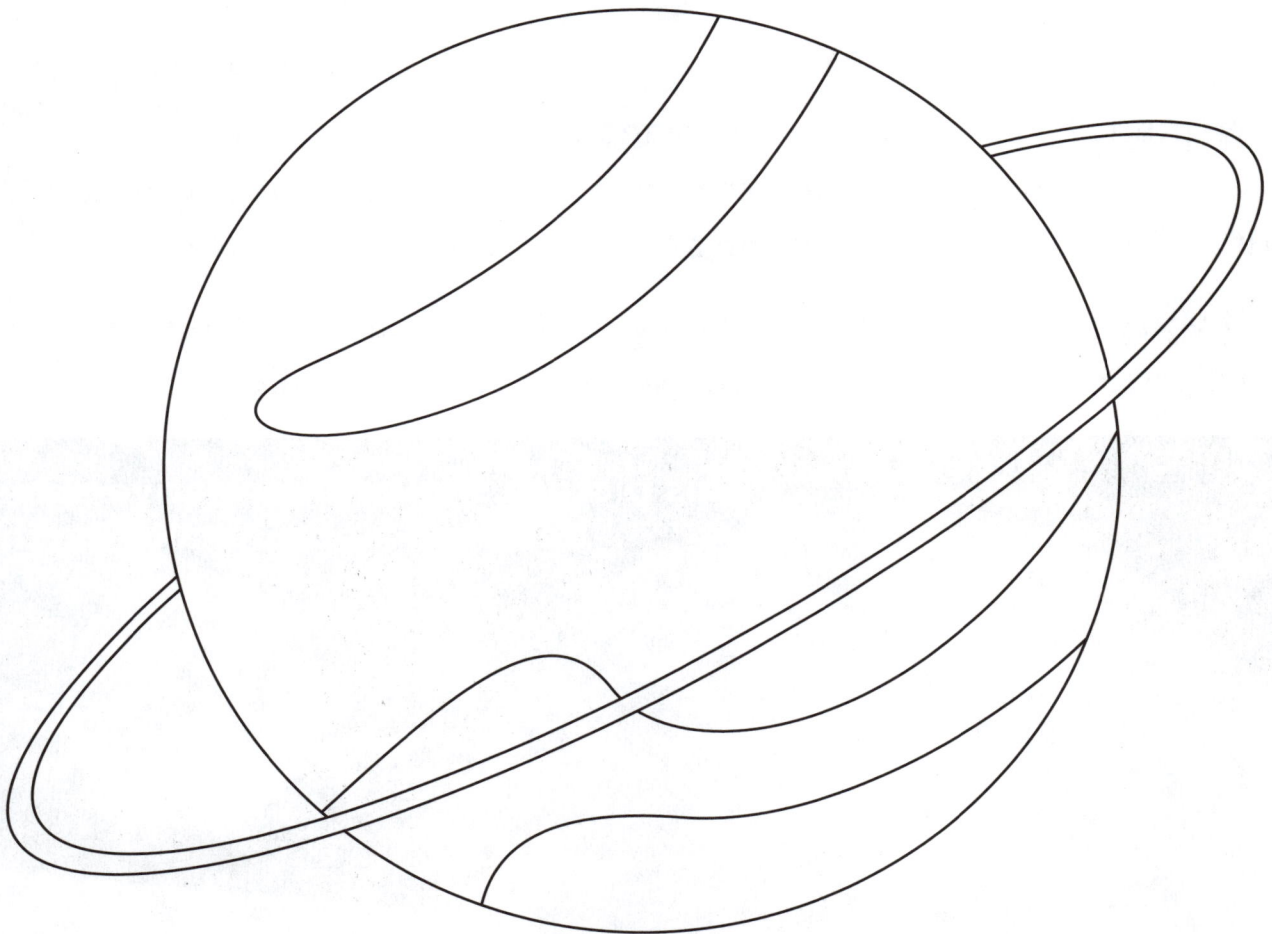

Uranus is the S_____ planet from the sun.

Uranus is a pale b_____ color.

Uranus is a g_____ planet.

Uranus has a few r_____ and at least 20 moons.

Neptune

Neptune is the eighth planet from the sun. It is about the same size as Uranus. Neptune is made of gas. Its atmosphere makes it look blue.

Neptune has at least four rings and fourteen moons. Two of the moons are large and the rest are small. The largest moon is called Triton. Triton is about the same size as our moon.

If you could go to Neptune, you would be so far away from the sun that the sun would appear as only a bright star in the sky.

This picture shows the relative sizes between Neptune and Earth.

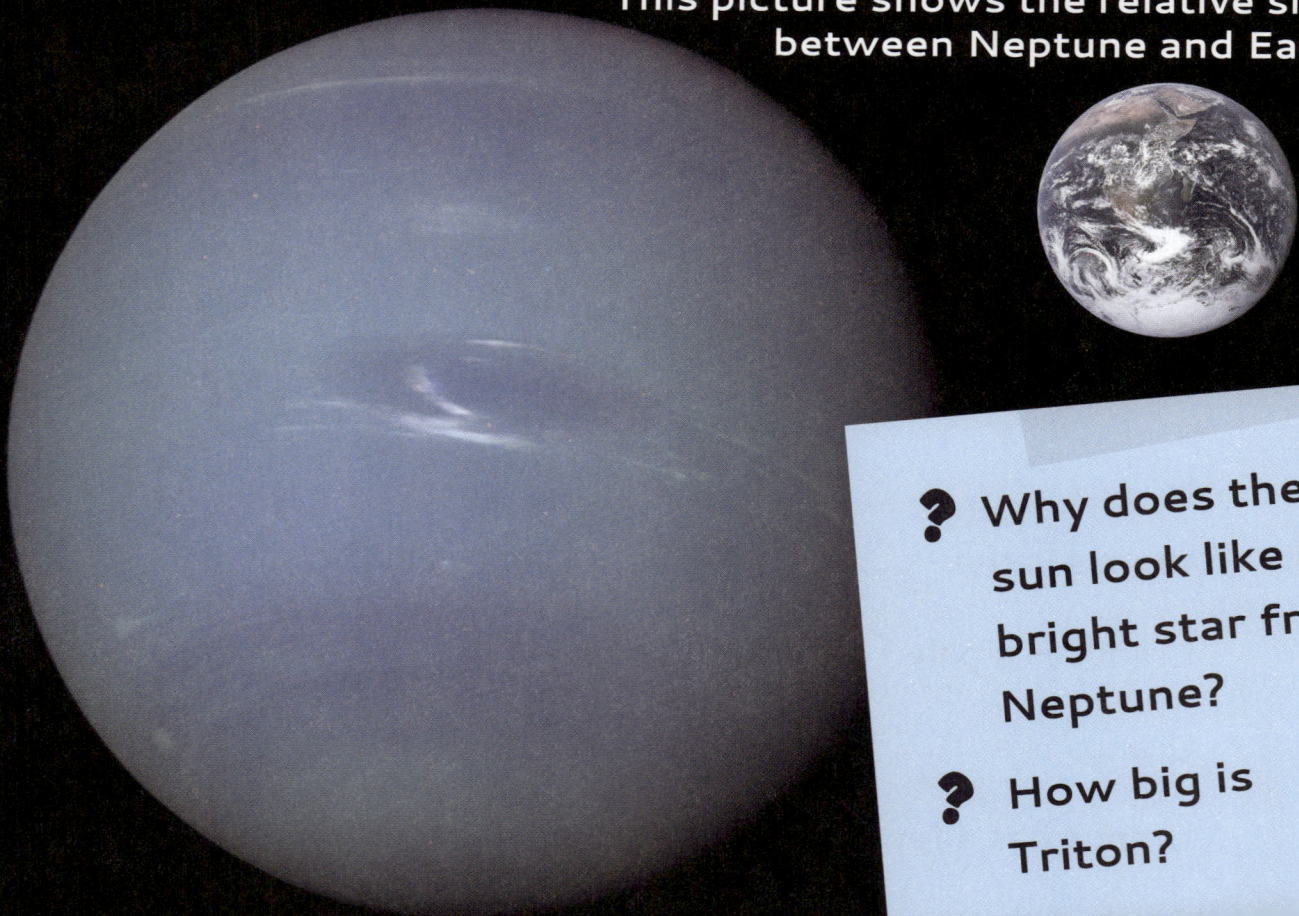

? Why does the sun look like a bright star from Neptune?

? How big is Triton?

Neptune Facts Sheet

Trace the name and color the planet. Review the facts you have learned about this planet and fill in the words below. The first letter of the answer has been given to you.

Neptune

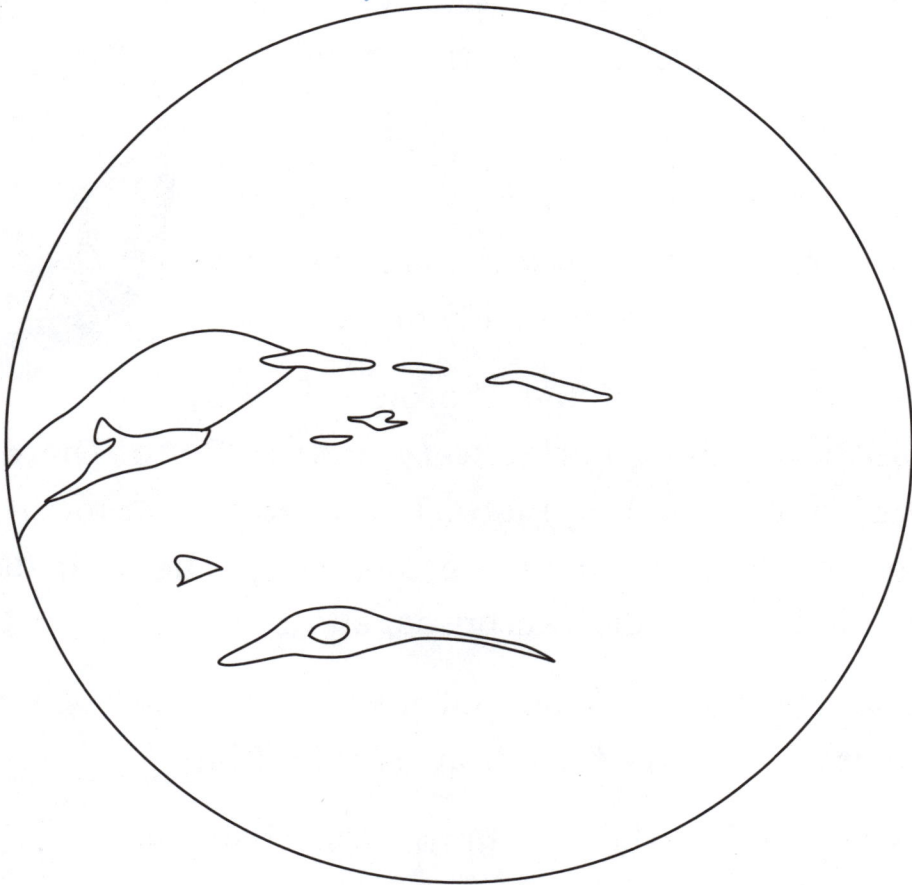

Neptune is the e_____ planet from the sun.

Neptune is about the same size as U_____ .

Neptune looks b_____ , like Uranus.

Neptune is a g_____ planet.

T_____ is Neptune's largest moon.

Pluto and Eris

Pluto and Eris are round objects that orbit very far from the sun. Because they are much smaller than regular planets, they are called dwarf planets. Pluto and Eris are very cold all the time. The sun is not close enough to give them any heat.

Pluto

Pluto has a large moon called Charon. It is about half as big as Pluto. Charon orbits very close to Pluto. There are four other moons orbiting Pluto. These moons are much smaller than Charon. And they orbit much farther away. These small moons have been named Nix, Hydra, Kerberos, and Styx.

Eris is about the same size as Pluto. Eris is very far away from the sun. It is about three times farther away than Pluto is.

Scientists are constantly looking for new objects in outer space. Scientists believe there are at least two other dwarf planets. They are

✏ Scripture Trace

Let the heavens praise Your wonders, O Lord. Psalm 89:5

farther away from the sun than Pluto is. Scientists don't know much about them yet.

Now you have learned about all of the planets and dwarf planets in our solar system! Review their names by singing the song you learned in lesson 11.

? **Why are Pluto and Eris called dwarf planets rather than just planets?**

? **Why are Pluto and Eris very cold?**

? **Which dwarf planet is farthest from the sun?**

Planet Review

You will need a tape measure and 10 blocks or other objects to use as markers. To appreciate how far Pluto and Eris are from the sun, do the following activity. If you only have a small space in which to do this activity, you can cut the distances in half and still get a good feel for how far each planet is from the sun.

Choose an object like a chair to be the sun. Set a block 2 inches from the chair. This represents how close "Mercury" is to the sun. Set a block 4 inches from the "sun" to represent "Venus." Place "Earth" 6 inches from the "sun" and "Mars" 8 inches away. "Jupiter" goes 2 feet and "Saturn" 4 feet from the "sun." Place "Uranus" 8 feet from the "sun." You may need to move out into a hallway or into another room to place the remaining blocks. "Neptune" should be 13 feet away and "Pluto" should be 17 feet away. Finally place "Eris" 41 feet away from the "sun." You may need to go outside to show how far away Pluto and Eris are from the sun. It is probably pretty hard to see the "sun" from your location at "Eris."

Unit Vocabulary Review

Review the names of the planets by writing each name in the space provided. The first letter of each name is given for you.

M _ _ _ _ _ _ _

V _ _ _ _ _

S _ _ _ _ _ _

E _ _ _ _ _

M _ _ _ _

U _ _ _ _ _ _

J _ _ _ _ _ _ _

N _ _ _ _ _ _ _

Write the names of two dwarf planets that orbit very far from the sun.

P _ _ _ _ _ and E _ _ _ _

Space Program

Universe
for Beginners

Lessons 28–35

GOD'S DESIGN®

National Aeronautics and Space Administration (NASA)

NASA is a group of people who work on many different things that help us understand space. The people at NASA plan and build spaceships. They built the Lunar Module and the Space Shuttle. They make rockets and space probes that visit other planets.

NASA trains astronauts to work in space. They helped build the International Space Station. And they plan many experiments for astronauts to do on the space station. Some people at NASA helped build and now operate the Hubble Space Telescope. This telescope helps us to see things in space that are very far away. The work that NASA does is very important. It helps us explore space.

Hubble Space Telescope

Hubble Telescope

With a parent's permission, visit the Hubble Telescope website. You can see what the telescope is looking at right now!

? **What is NASA?**

? **Name three types of things that NASA does.**

? **What is the name of the special telescope that NASA built in space?**

What Does NASA Do?

NASA does things that help us understand space.

✗ Mark out the pictures that do not have to do with learning more about space.

Planets and solar system

Doctor

Astronaut

Dog

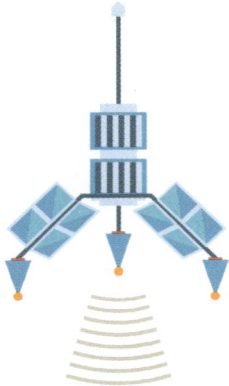

Space probe

House

Space Station

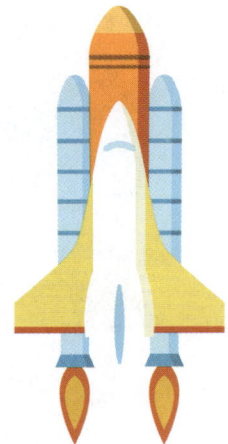

Space shuttle

Moon Rover

Car

Telescope

Rocket about to be launched

Space Exploration

In order to send something into space you must have a rocket. A

rocket is a very powerful engine. It can push something up into the air very quickly. The first thing that was put into space was a Soviet satellite called

Sputnik. A *satellite* is something that orbits the earth. Rockets have also been used to put men into space. The first man to go into space was Yuri Gagarin, from Russia. Rockets have been used to send men to the moon. The first people to visit the moon were the Americans Neil Armstrong and Buzz Aldrin.

Rockets send other things into space, too. They were used to send the parts into space to build the Space Station. It took many trips to get all of the parts into space. Once the parts were there, astronauts went up into space in the space shuttle. There they put the parts together.

Space probes

have been sent into space too. A space probe is an instrument that visits far away places and planets. It usually has cameras and scientific tools. Probes send information back to Earth so we can learn about things that are very far away.

Launching Rockets

Rockets are needed to launch things into space. A rocket works because hot gas rushes out the back of the engine. This pushes the rocket forward. To see how this works, blow up a balloon but do not tie it shut. When you let go of the balloon, what happens? Air rushes out of the balloon and this pushes the balloon forward. Have fun turning your balloon into a rocket engine!

? What is a rocket?

? What is a satellite?

? What are some things that have been sent into space by rockets?

Space Probe Paths Worksheet

Trace the paths in the solar system to see where space probes have gone from Earth.

Mars

Ceres

Saturn

Moon

Pluto

Titan

Mercury

Neptune

Venus

Uranus

Jupiter

Apollo Program

It was not an easy thing to send a man to the moon. No one had ever been in space before. But NASA was able to do it. First, they sent one man into space. Next, they sent two men into space so they could learn how to work together. After several flights in space near Earth, they were ready to send people to the moon. The project to send people to the moon was called the

Apollo Program.

Scientists first had to design a rocket that was big enough to send people that far. The rocket was

called the Saturn V (five). It was a very big rocket. It was really three rockets in one. The first rocket fired and lifted the astronauts into space. When it ran out of fuel, it dropped off

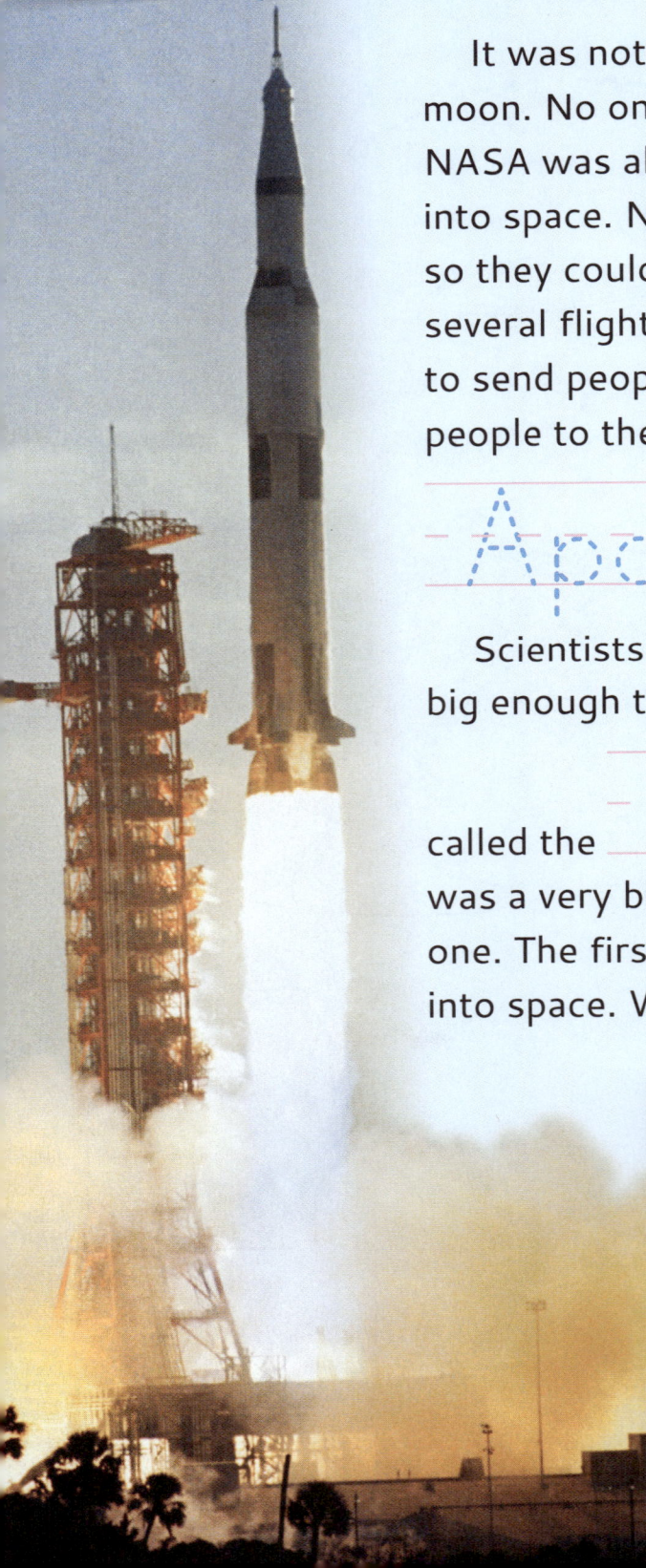

Apollo command capsule

and the second rocket fired. After it dropped off, the third engine fired and sent the men all the way to the moon.

The space capsule sat on top of the rocket. It had a place for the astronauts to sit and a place for food, computers, and other instruments. It also carried the Lunar Module. The _Lunar Module_ took Neil Armstrong and Buzz Aldrin down to the moon.

The astronauts took samples of moon rocks and moon dust. They did many experiments on the moon. They explored the moon for a little more than two hours. When they were done, the Lunar Module took them back to the space capsule, and they returned to Earth.

Lunar Module

Lunar Rover

Astronauts made a total of six trips to the moon. On later trips to the moon, the astronauts took a vehicle called the ⸻Lunar Rover⸻ with them. This "moon car" helped the astronauts explore the moon's surface.

Lunar Rover

Make an edible lunar rover. Use sandwich cookies or round crackers as the wheels. Use a graham cracker as the base. Use peanut butter or sunflower seed butter to stick everything together.

Imagine what it might have been like to walk on the moon with Neil Armstrong. While you eat your lunar rover, watch the video of the Apollo 8 astronauts reading from the book of Genesis (a parent can find this on the NASA website).

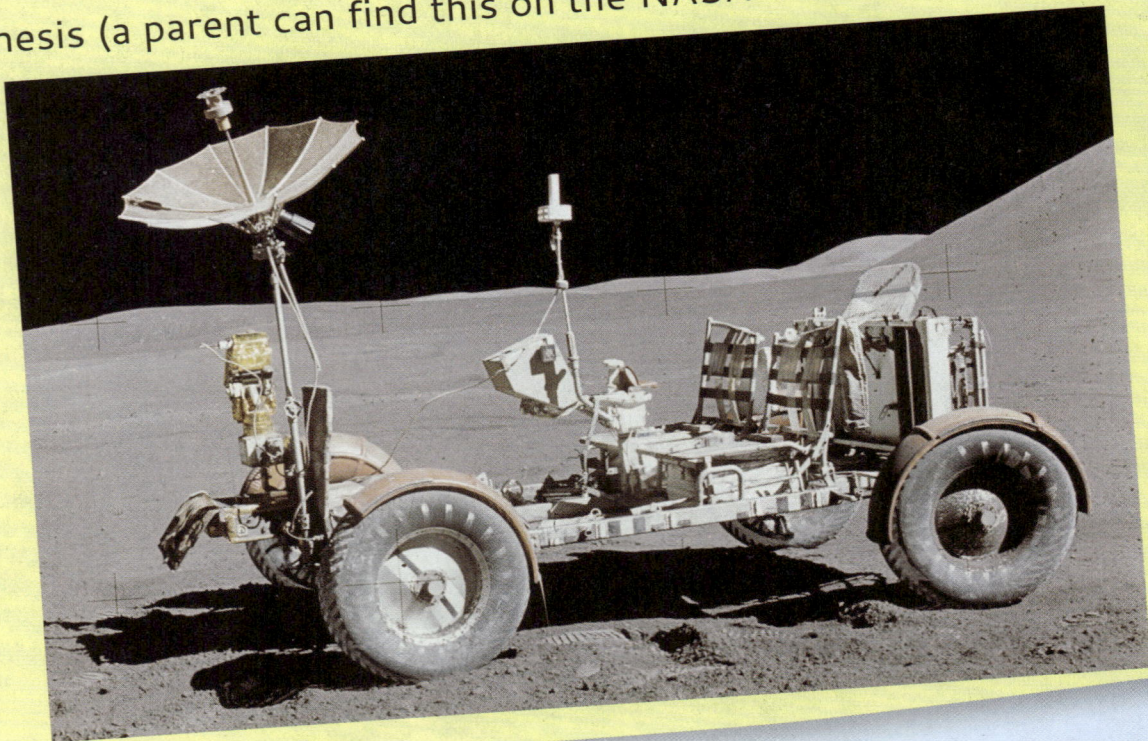

Moon Rover Coloring Sheet

Color this picture of astronauts exploring the moon.

? What was the Saturn V?

? What was the job of the Lunar Module?

? What vehicle was used to explore the moon?

The Space Shuttle

Only a few astronauts have gone to the moon. Most people who have gone into space have stayed close to Earth. The

space shuttle was designed to take people into space close to Earth. The space shuttle system had three parts. The orbiter looked like an airplane without jet engines. It had triangle-shaped wings and a tail fin. The second part was a big orange fuel tank that held liquid fuel. The third part consisted of two rocket boosters which held solid fuel.

The orbiter had a special place where the astronauts lived while on a mission. It also had a place to store satellites or other equipment. This bay had a robotic arm that was very useful for launching satellites. The arm was also used for building and repairing the space station.

Crew compartment

Storage compartment

Robot arm

Discovery

(Right) Here is an astronaut using the shuttle's robot arm to help him do repairs in space.

Astronauts did many different things on the space shuttle. Sometimes they did experiments. Other times they launched satellites. Many times, they took supplies and people to the International Space Station. When a mission was complete, the space shuttle returned to Earth. It landed in either Florida or California. The final space shuttle mission ended on July 21, 2011.

? What was the space shuttle designed to do?

? What are the three parts of the space shuttle?

Space Shuttle Coloring Sheet

The crew of the shuttle lived and worked in the front part of the orbiter. The big area in the middle was where they stored things they were taking into space such as satellites or supplies. Draw in some supplies for the astronauts to take into space. They would need to take food, water, spare parts, equipment for science experiments, and extra air in case of an emergency. Draw a long robot arm in this area too. There is a picture of one on the previous page. This arm was used to put satellites into space. Now color your picture.

International
Space Station

The _International Space Station_ was built so that astronauts could stay in space for many months. Many countries have worked together to build the space station. The United States, Russia, Canada, Japan, and several nations in Europe have taken part in this huge project.

The space station has several parts. It has solar panels that provide energy for the space station. There are crew quarters where the astronauts live. There are several laboratories. This is where the astronauts do experiments.

There are usually three astronauts living on the space station at a time. Most of the astronauts have been from either the United States or from Russia. But several have been from other countries as well.

On the space station, astronauts can do experiments that cannot be done on Earth. There is very little gravity on the space station. So, things act differently up there. Astronauts experiment with plants and animals. They test chemicals and medicines. Astronauts have learned many helpful things in space.

? **What is the purpose of the International Space Station?**

? **How does the space station get power?**

Water Balls

Since there is very little gravity on the space station, water acts differently there than it does on Earth. Water forms into balls. This is because water drops like to stick together. You can see water drops stick together on Earth, too. Try this experiment:

1. Place a few drops of water about 2 inches apart on a piece of waxed paper. Notice how the drops form into bubble shapes. They do not flow out across the paper. On the space station these drops of water would form balls. On Earth, gravity flattens them into bubbles.

2. Use the edge of a butter knife to slowly push one drop of water toward another. When they get close, they should pull together to form a bigger bubble of water. This is because water drops want to be close together.

3. Try to separate the water bubble into two smaller bubbles. This will be very hard because the water wants to stay together.

To see how water really acts on the space station, ask your parent to find a video on the internet of an astronaut wringing out a wet washcloth on the space station.

Design Your Own Space Station

Imagine you work for NASA and they want you to draw a new international space station. Draw your own idea of what a great space station should be. Make sure to include these parts in your drawing:

- ◆ **docking port** (where visitors attach their space vehicles to go inside)
- ◆ **storage** (for food and other essentials)
- ◆ **living areas** (kitchen, bedrooms, and more)
- ◆ **lab** (where the astronauts can work)
- ◆ **solar panels** (for the station to have power)

Astronauts

Have you thought about becoming an astronaut?

Astronauts

are people who work in space. If you want to be an astronaut, you have to learn math and science. And you have to be in good shape. Then, when you are an adult, you can apply to the space program. Many people apply, but only a few are chosen to become astronauts.

An astronaut has to learn many new things. Astronauts must learn how to work without gravity. They

must learn to use special tools and machines. They might have to learn to fly a spaceship.

Much of the work that is done in space can be done inside the space station. There is air in the station. But sometimes work has to be done in space, where there is no air. Astronauts have to wear special

spacesuits to keep them alive in space. The suits supply them with air and water. Spacesuits also provide heat and cooling. They even have radios. This lets the astronauts talk with each other. They can talk with people back on Earth, too. Sometimes astronauts put on rocket packs that help them move around in space.

You will have to work very hard if you want to be an astronaut. But it will be a very special job if you make it.

Wear a Spacesuit

Pretend you are an astronaut and you must wear a spacesuit. Use your winter coat and snow or wind pants as the main part of your spacesuit. Add snow boots and heavy gloves to protect your hands and feet. Finally, put on a helmet. You can use your bike helmet.

Now you are dressed in your spacesuit. Try to do some of the activities that an astronaut might do. Try to build something with blocks. Try to screw a nut onto a bolt. It is difficult to do these things with heavy gloves on. Astronauts must learn to work with thick gloves on their hands. They often have special tools to help them do their job.

Do some exercises in your spacesuit. Did you get too hot? Real spacesuits have a way to keep the astronauts cool when they are working hard in space.

? What are two things you need to study in school if you want to be an astronaut?

? What are three things that a spacesuit provides that are missing in space?

Astronaut Dot-to-Dot

Connect the dots to see what an astronaut looks like in space.
Color and draw things that are seen in space around the astronaut.

Final Project— Solar System Model

Now that you have learned about space, you can make your own model of the solar system. Use modeling clay to make the sun and the eight planets. Look back at the vocabulary review after Lesson 27 to see how big to make each planet compared to the sun. If you have different colors of clay, you can look at the previous lessons to see what colors to use for each planet. Place the planets in the right order on the table to finish your model. As an alternative to making a clay model of the solar system, you can purchase inexpensive Styrofoam™ or plastic models that you can paint.

? Which is your favorite planet?

? What did you like learning the most in this book?

✏️ **Scripture Trace**

Lift up your eyes on high and see: who created these?
Isaiah 40:26

Conclusion

God has created an amazing universe for us to live in. Think about all the things you have learned about in this book. God made the sun, the moon, and all the stars. He made the planets. And best of all, God made Earth for us to live on. Tell God thank you for making such a wonderful place for us to live.

Viewing the Stars

Go outside at night and find a dark place to spread a blanket on the ground. Lie on your back and watch the stars. Think about what it would be like to be able to visit all of those faraway places. Look for pictures in the stars (constellations). See if you can find the Big Dipper and the Little Dipper again. You can also look online (with your parent's permission) for different views of the stars.

Scripture Trace

Height nor depth, nor anything else in all creation, will be able to separate us from the love of God in Christ Jesus our Lord.   Romans 8:39

Planet or Not a Planet Worksheet

Write each word on the correct side of the page. If it is an actual planet, write it on the left side. If it is not a planet, write it on the right side.

Mercury   Comet   Earth   Sun
Moon   Pluto   Venus   Meteor
Asteroid   Mars   Jupiter   Titan
Saturn   Neptune   Star   Uranus

Planet	Not a Planet

Unit Vocabulary Review

Use the clues and the bank to complete the crossword puzzle.

Apollo   astronaut   module
NASA   rocket   rover
satellite   Saturn V   shuttle
spacesuit   station

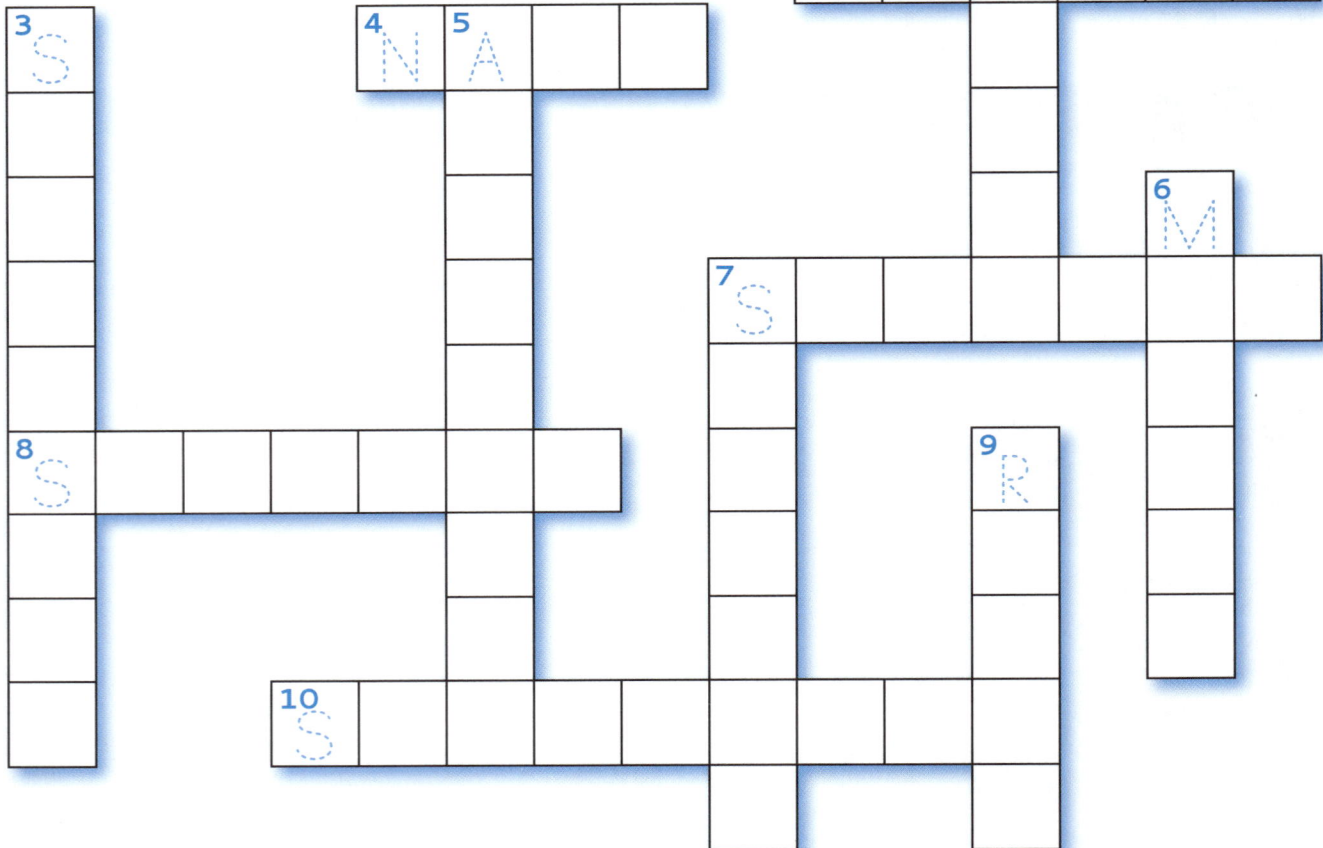

Across

2. Program that sent men to the moon
4. Group of people who build spaceships
7. The International Space _____ allows people to live in space for months.
8. Rocket that sent men to the moon
10. Something that orbits the earth

Down

1. Powerful engine that sends things into space
3. Must be worn when working in space
5. Person who works in space
6. Vehicle that lifted men off of the moon was the Lunar _____.
7. The Space _____ was designed to take people into space close to earth.
9. The "moon car" that helped men to explore the moon was the Lunar _____.

Planet Earth
for Beginners

GOD'S DESIGN®

Origins and Glaciers

UNIT
1

Planet Earth
for Beginners

GOD'S
DESIGN®

Lessons 1-7

Introduction to Earth Science

We all live on a planet called Earth.

Earth science

is learning all about the planet earth. There are many questions we can ask about the earth. Where do rocks come from? What is a cave? What makes a volcano erupt? We will learn the answers to these questions.

The most important thing you can know about the earth is that God created it. In the Bible, in Genesis 1:1, it says, "In the beginning God created the heavens and the earth." The Bible tells us that God also created the sun, moon, stars, sky, dry land, and every kind of plant and animal. So, as you learn about the earth, look for things that God made. You will find that He created a really wonderful place for us to live.

? **What is earth science?**

? **Where did the earth come from?**

? **What other things did God create?**

✏️ ## Scripture Trace

In the beginning, God created the heavens and the earth.

Genesis 1:1

Creation Scavenger Hunt

Go outside. Look all around you. You are on a hunt for things that God created. Your scavenger hunt worksheet has pictures of things God created. Check off each item as you see it. For more fun, take a sheet of paper and draw pictures of other things that God created that were not listed on this worksheet.

☐ Sun

☐ Plants

☐ Trees

☐ Squirrels

☐ Clouds

☐ Rabbits

☐ Mushrooms

☐ Cats

☐ Nuts and seeds

☐ Dogs

☐ Bees

☐ Birds

☐ Butterflies

☐ Frogs and lizards

The Earth Is Special

The earth is a special place. Most of our planet is covered

with **water**. People, animals and plants all need
water. No other planet has water like earth does. God placed

the earth just the right distance from the **sun**.
The sun keeps our planet warm. But we are not too close.
Planets that are closer to the sun are too hot to live on.

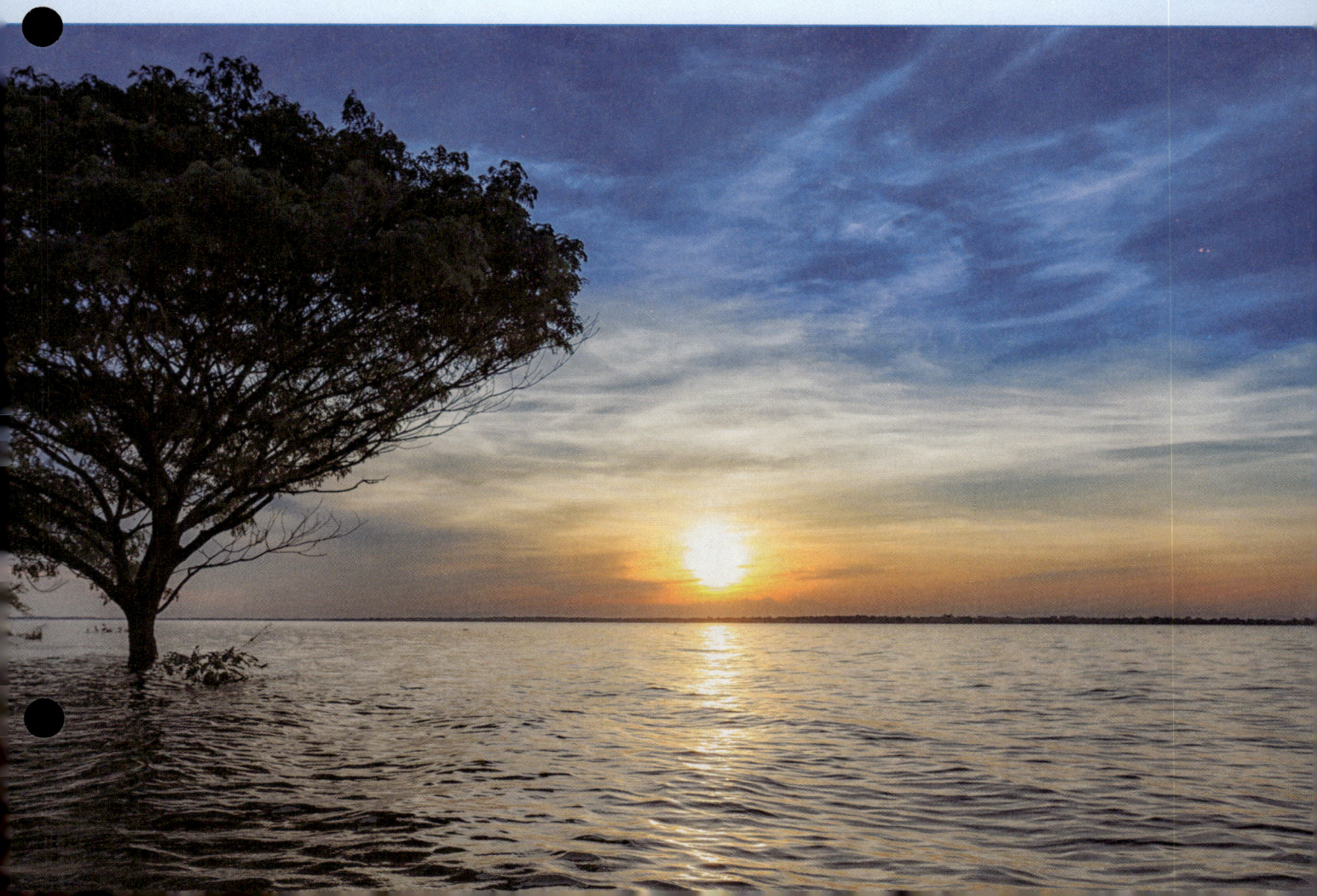

God also put many important things in the ground when he made the earth. The land has soil so that plants can grow to provide us food, medicine, and wood for building. There is metal in the ground. People use metal to make cars and other machines. There is oil in the ground. People use oil to make plastic and gasoline. Everything you use comes from our planet. God made Earth special.

? **What are some things in your house that are made of metal?**

? **What are some things in your house that are made of plastic?**

? **What are some things in your house that are made of wood?**

? **Why is Earth a special planet?**

Earth Products Scavenger Hunt

Make a chart for your wall with three columns labeled Metal, Plastic, and Wood. Now look around your house for items that are made from metal, plastic, and wood. Draw a picture of each item or write its name in the correct column.

Earth by Number

Color the picture of our special planet using the addition facts.
 If the numbers add up to 4, color that space blue.
 If the numbers add up to 6, color the space green.
After you color the picture, trace the words below it.

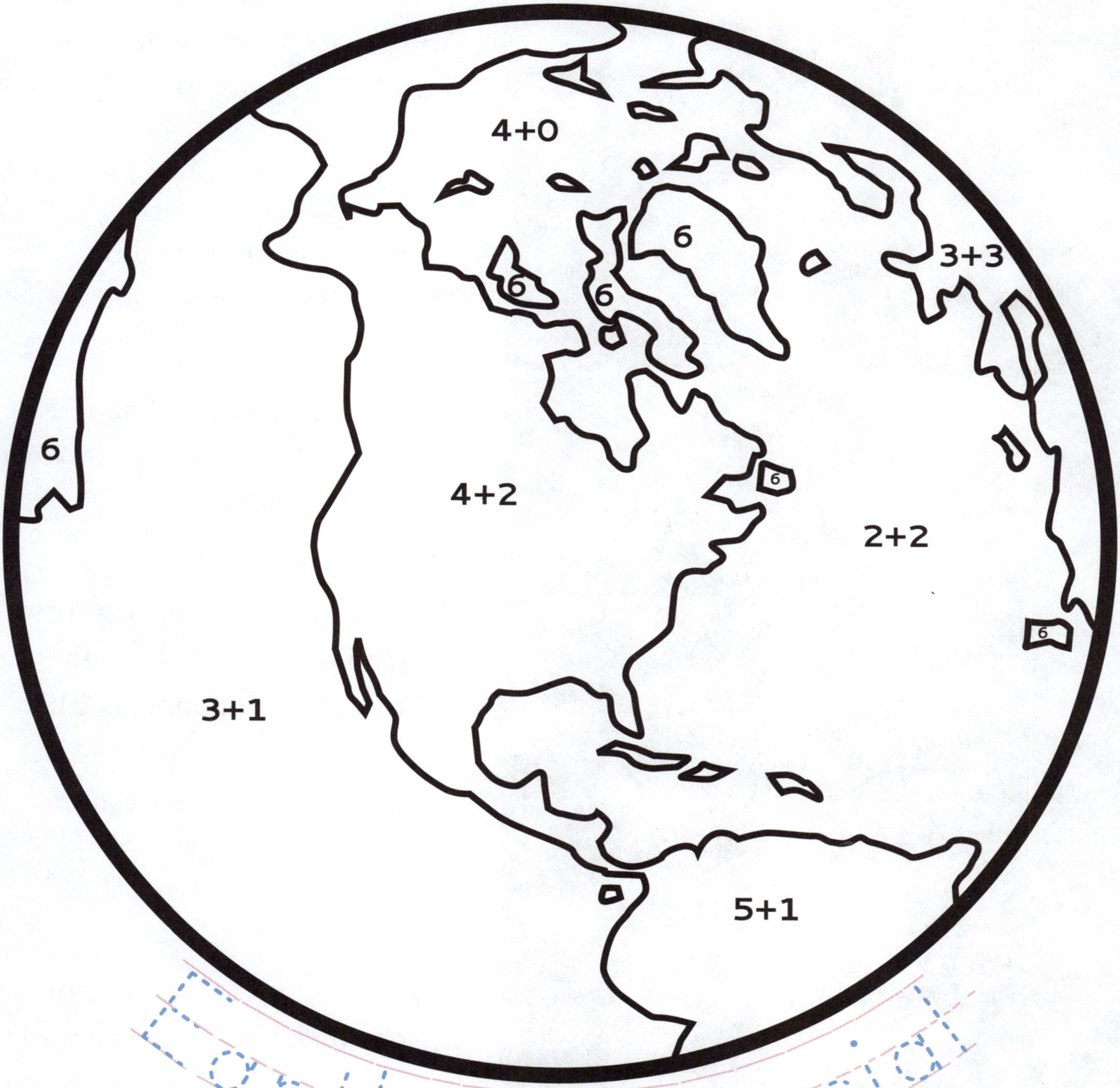

4+0

6

3+3

6

6

6

4+2

6

2+2

6

3+1

5+1

Earth Is Special

The Earth's History

The Bible tells us that God created the earth about 6,000 years ago. This

creation

happened in only six days. The earth was very good when God made it.

God also created Adam and Eve, who were the first man and woman.

But Adam and Eve disobeyed God. They ate from the tree that God told them not to eat from (Genesis 3). This is called the

fall of

man

. As part of their punishment, God cursed the earth.

The earth is different from the way God originally created it.

The earth now produces weeds and thorns. This makes it difficult to grow the plants that we want.

By the time Noah was alive, (Genesis 6–8) the people had become very wicked. God sent a great

flood to punish them. Water covered the whole earth. This again changed the way the earth looks.

Some people do not believe the Bible and think that the earth is millions of years old. But when we learn about fossils, rocks, and other things on earth, we will see that they agree with what the Bible says and show that God's Word is true.

? What three events did you learn about that have changed the earth?

? Does the Bible tell us the earth is young or old (thousands or millions of years)?

✏️ Scripture Trace

Cursed is the ground because of you; . . . thorns and thistles it shall bring forth for you.

Genesis 3:17–18

Events Worksheet

These are major events in the history of the earth.

Write **1** in the circle next to the picture of creation.

Write **2** in the circle next to the picture of Adam and Eve disobeying God.

Write **3** in the circle next to the picture of God punishing all the wicked people during Noah's time.

The Genesis Flood

When Noah was alive most people were very evil. This made God very sad. He had to get rid of all of the evil. So God decided to send a flood that would cover the whole world. But Noah loved God. So, God saved Noah, his wife, their three sons, and their sons' wives. God told Noah to build a giant boat. This boat is called

the __ark__.

God sent at least two of every kind of land animal and bird to Noah so they could be saved on the ark. All of the animals (including dinosaurs and flying reptiles like pteradons) went into the ark. Noah's family also went into the ark. Then God shut the door. God sent a great flood that covered the entire world. All of the wicked people died. All of the animals that lived on the land died. But the people and animals on the ark were saved.

A fossil is formed

Flood in a Jar

Put a handful of sand in the bottom of a large jar. Add a handful of dirt on top of the sand. Add a handful of pebbles then a few rocks on top. Add a few twigs and dried leaves. The jar should be about half full. These are the types of things that were on the earth before the Flood.

Fill the jar most of the way with water. Put the lid on the jar. Shake the jar for 30 seconds. Set the jar down. The water will be filled with mud and dirt. This is how the floodwaters looked during the Flood.

Wait 30 minutes, then look at the jar again. You should see layers forming in the bottom of the jar. Many of the rock layers were formed around the world as particles in the waters of the Flood settled out into layers.

Did any of the twigs get buried by the mud? Animals and plants were buried in the mud during the Flood. This is how animals and plants became fossils after the Flood.

The floodwaters carried mud and dirt with them. This mud covered many of the animals. Some of these animals eventually turned into fossils. Layers of dirt and mud piled up. Much of this dirt and mud turned into rocks. In other areas, the water washed away rock and dirt. This changed the way the earth looks. When we study the earth today, we see many results of this great flood.

? Why did God send the Great Flood?

? What is the ark?

? How did the Flood change the surface of the earth?

? Were there dinosaurs on the ark?

After the Great Flood

Connect the dots to reveal the ark. Then count the animals coming off the ark and color the picture.

Total pairs:
(count by twos)

Total animals:

The Great Ice Age

Before the Genesis Flood, the weather was warm over most of the earth. After the Flood, things were different. The land was cooler. There were more clouds. There was ash in the air because of volcanoes. It began to snow in many areas. In some areas the snow did not melt, even during the summer. The ice began to build up in these areas. So much snow and ice formed in parts of

the world that it was called the __Ice Age__.

During the Ice Age, many parts of the world were covered with ice. But other areas were still warm. People and animals lived in these warmer areas. The Ice Age lasted for several hundred years. The land eventually warmed up. Finally, the climate (weather over a long time) became much like it is today.

Ice That Won't Melt

You need two bowls. Put an ice cube in each bowl. Place one bowl in the refrigerator. Place the other bowl on the counter. It is warm on the counter. This is how the weather was before the Flood. It is cold in the refrigerator. This is how the weather was in some areas after the Flood. Which ice cube will melt the fastest? Wait 30 minutes. See which ice cube has melted the most. (If you have access to a small scale or balance, you could weigh the ice cubes before you put one in the refrigerator and again after 30 minutes to see how much they change.)

When the weather is cold, the ice does not melt very quickly. This is how ice lasted for hundreds of years during the Ice Age.

? How were things different on the earth after the Flood?

? Why did the ice last so long during the Ice Age?

Ice Age Animals

During the Ice Age, many animals had long fur to protect them from the cold.

✗ Mark out the animals that would not live in the snowy areas during the Ice Age.

Find the name of each animal at the bottom of the page. Draw a line from each animal to its name.

elk   frog   mammoth   toucan

arctic fox   lizard   snake

Planet Earth Lesson 5   **263**

Glaciers

Glaciers are thick sheets of ice. The ice in a glacier never completely melts. Snow falls during the winter. During the summer, some, but not all, of the snow melts. The next winter, more snow falls. The new snow presses the older snow down. This turns it into ice. Each summer, some of the ice melts. Each winter more ice is added. Many of the glaciers that exist today have been around since the Ice Age. Most glaciers today are found near the north and south poles. You can look at a globe to see where the North and South Poles are.

Sometimes glaciers reach the water. When this happens, chunks of the glacier break off. Ice is less dense (lighter) than water, so these chunks float in the water. These floating chunks of ice are called

icebergs.

Icebergs can be small or very large. When they are large, most of the iceberg stays below the surface of the water. Only part of the iceberg can be seen above the water. This can be dangerous for ships. If a ship sails too close to an iceberg, it could hit the part that is underwater. In 1912, a ship called the *Titanic* hit an iceberg and sank.

Tiny Icebergs

Fill a glass container with water. Add a few ice cubes. The ice cubes are like tiny icebergs. Look at the ice cubes through the side of the container. How much of the ice cube is above the surface? How much is below the surface? Icebergs that break off of glaciers are floating in the ocean like these ice cubes. A large part of the iceberg is below the surface. It cannot be easily seen.

? **What is a glacier?**

? **Where are most glaciers found today?**

? **What is an iceberg?**

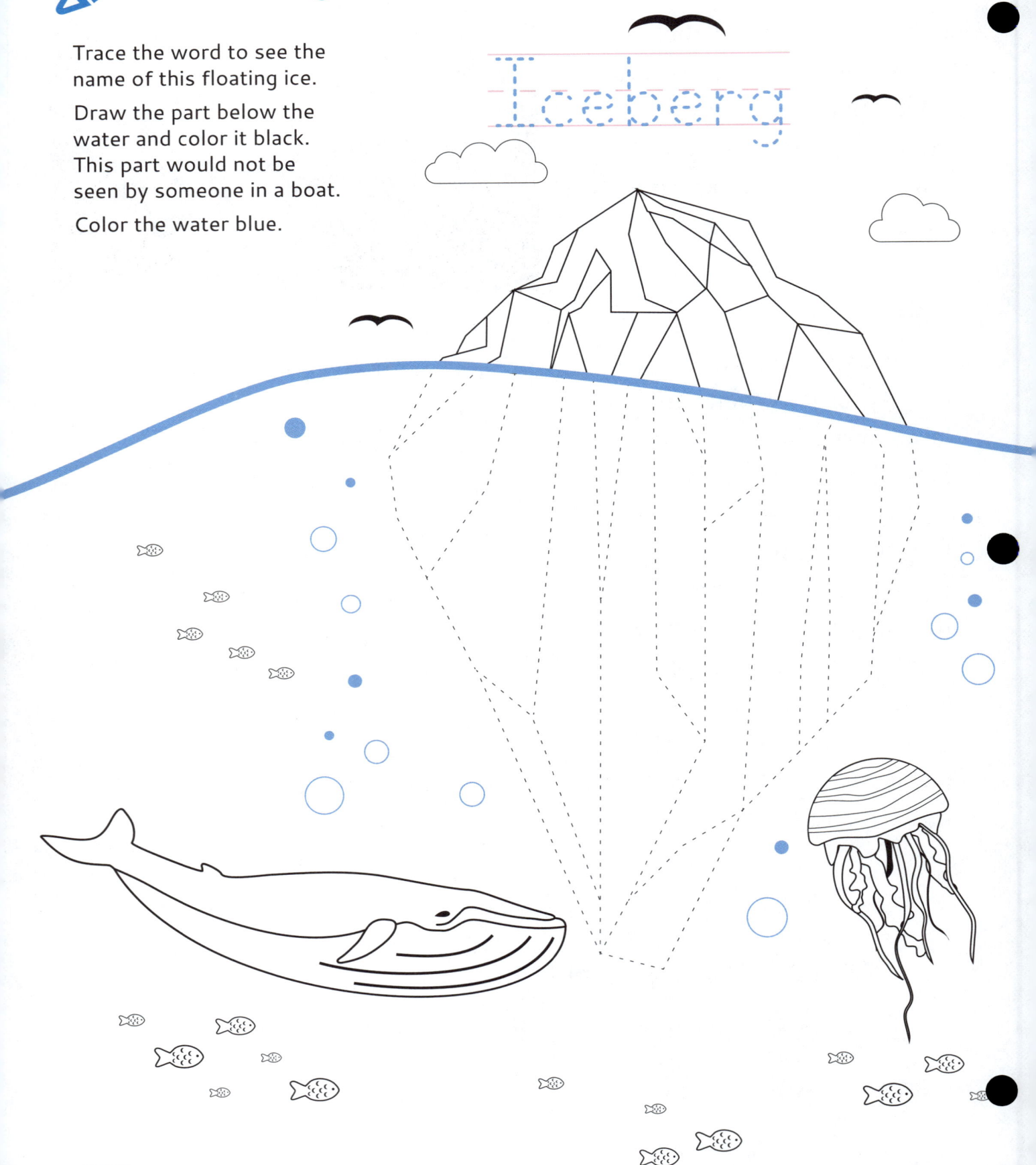

Floating Ice

Trace the word to see the name of this floating ice.

Draw the part below the water and color it black. This part would not be seen by someone in a boat.

Color the water blue.

Iceberg

Glaciers are large sheets of ice. Glaciers don't just stay in one place.

They can move. Gravity is the force that pulls everything down toward the earth.

Gravity pulls on glaciers. This makes the ice move slowly down a mountain. When the ice moves, it drags rocks and dirt with it. When the weather is warm, the ice begins to melt. Water flows into the soil and rocks under the glacier. When the weather becomes cold, this water freezes. The rocks become frozen in the ice. When the ice is pulled down a hill by gravity, the rocks in the ice act like sandpaper. They scrape the ground underneath the glacier.

Glaciers also push rocks and dirt ahead of them. When the glacier melts, it leaves a line of rocks showing where the ice has been. Glaciers usually move very slowly. They move only a few inches each day.

- **?** What makes glaciers move down a mountain?
- **?** How do rocks get caught in a glacier?
- **?** What do glaciers push in front of them?
- **?** Do glaciers move quickly or slowly?

Caught in a Glacier

Place a few pebbles or some sand in the bottom of a bowl. Set several ice cubes on top of the pebbles. Leave the bowl sitting on the counter for a few minutes. The ice will begin to melt. This is what happens to glaciers during the summer. Once there is some water in the bottom of the bowl, place it in the freezer. This is what happens during the winter. After one hour, take the bowl out of the freezer. Remove the ice cubes. You should see that the rocks have become frozen in your "glacier."

Scripture Trace

The waters become hard like stone, and the face of the deep is frozen.

Job 38:30

Glacier Maze

Find a way through the glacier from the mountains to the sea. Collect the letters along the correct path to discover a hidden word.

Put your collected letters in order here:

_____

_ _ _ _ _ _ _ _ _ _ _

Unit Vocabulary Review

Fill in each blank with one of the vocabulary words below.

1. E _____ S _____ is the study of our planet.

2. Two things that make Earth special are the W _____ that covers most of the surface and the distance we are from the

   S _____ .

3. Three major events that have changed the way the earth looks are

   C _____ , the F _____ of man, and the Great

   F _____ .

4. Noah built a very large boat which we call the a _____ .

5. During the I _____ A _____ many parts of the earth were covered with ice.

6. G _____ are large sheets of ice that do not completely melt in the summer.

7. I _____ are chunks of ice that fall off of glaciers and float in the water.

8. G _____ pulls glaciers down mountains.

ark	**fall**	**gravity**	**sun**
creation	**Flood**	**Ice Age**	**water**
Earth science	**glaciers**	**icebergs**	

Rocks and Minerals

UNIT
2

Planet Earth
for Beginners

Lessons 8-18

GOD'S
DESIGN®

Design of the Earth

The planet earth has three main parts. The **crust** is on the outside. This is the only part of the earth that we ever see. The crust is made of rock and is the thinnest part of the earth.

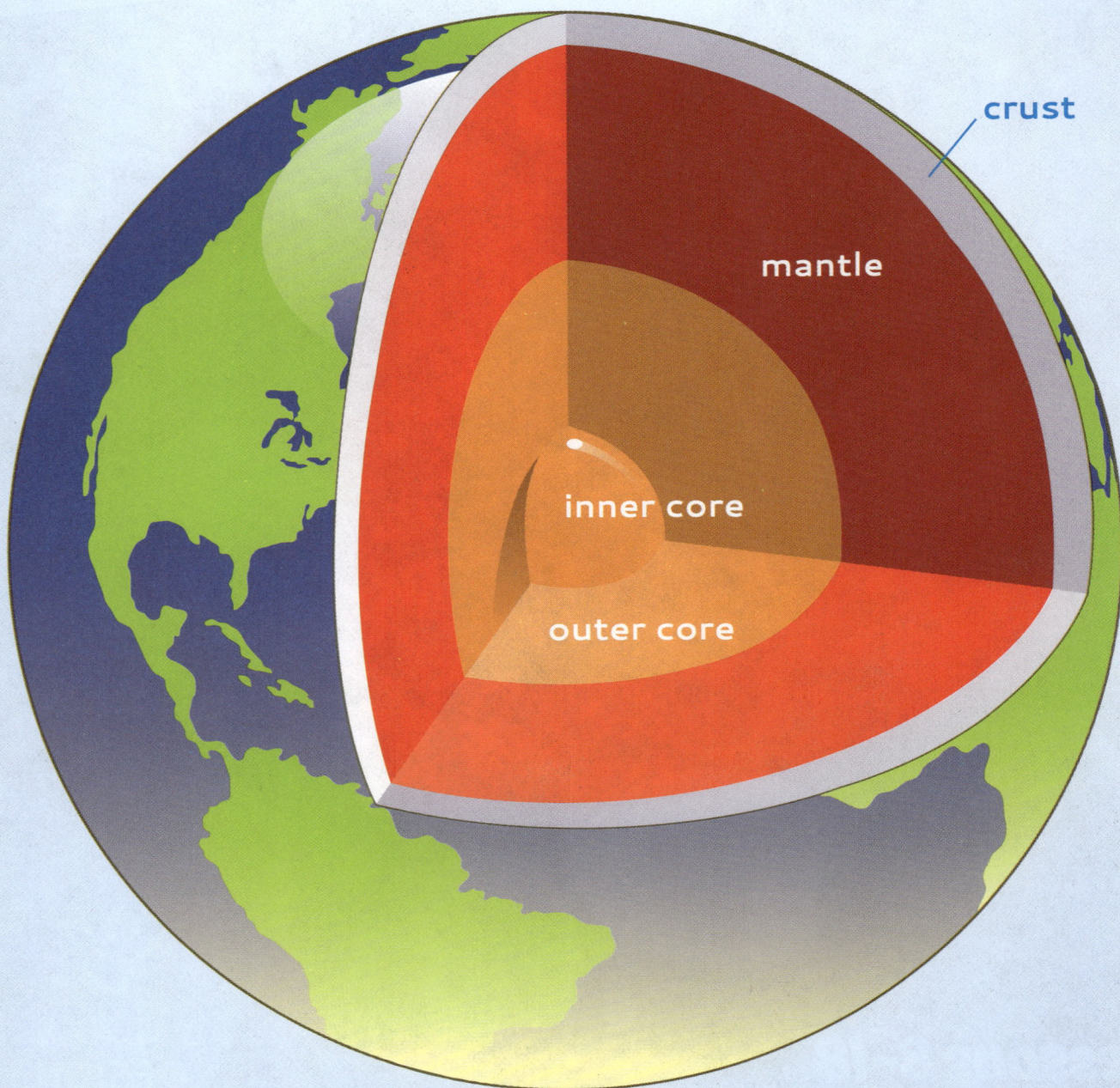

Underneath the crust is the _mantle_. The mantle is the thickest part of the earth. The mantle is very hot. Rocks inside the mantle can melt. Melted rocks inside the earth are called _magma_.

In the center of the earth is the _core_. The core is also very hot. It is even hotter than the mantle. The core has two parts. The outer core is liquid metal and the inner core is solid metal.

? What are the three main parts of the earth?

? Which part is the thinnest?

? Which part is the hottest?

Model of the Earth

An apple can be used as a model of the earth. With an adult's help, cut an apple in half. The skin of the apple represents the earth's crust. The flesh of the apple represents the earth's mantle. And the core of the apple is in the center of the apple just as the core of the earth is in the center of the earth. Now enjoy eating your yummy Earth model.

Center of the Earth Maze

Find your way into the center of the earth. Write the correct letter in each circle to show the parts of the earth.

Mantle=M   Crust=C   Inner core=I   Outer core=O

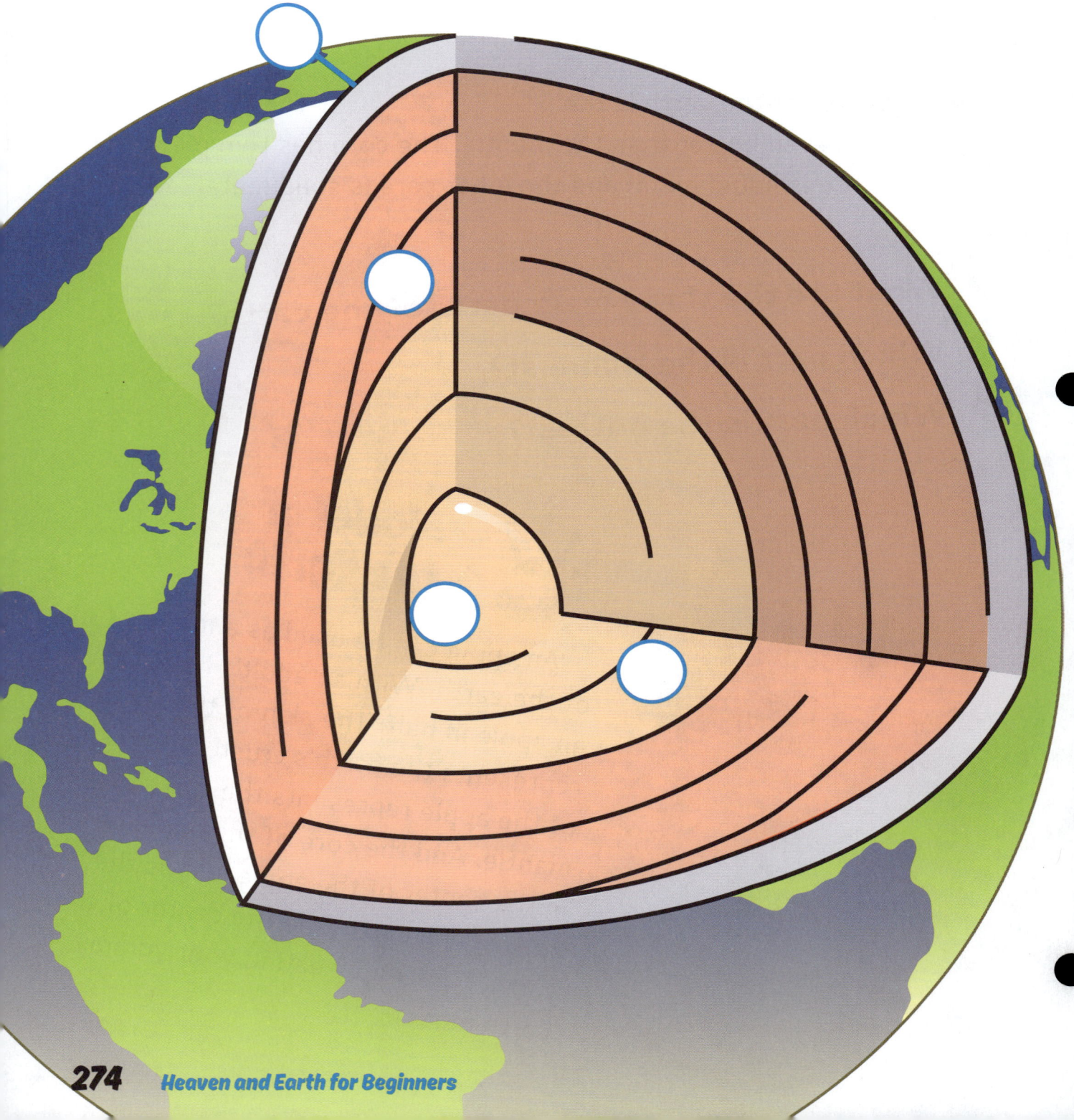

Rocks

Rocks are always underneath you, even when you are swimming in a lake or the ocean. There are rocks below the water. There are rocks below the sand. If you dig down below the dirt, you will eventually hit rock. These rocks form the crust of the earth.

There are three kinds of rocks. _Igneous_ rocks are formed when liquid rock (magma) cools. _Sedimentary_ rocks are formed when bits of sand and broken rocks are stuck together. The third kind of rock is metamorphic rock.

Igneous

Sedimentary

Metamorphic

Metamorphic rock is rock that has changed. Heat and pressure change it from one kind of rock into another.

We will learn more about each of these kinds of rocks in the next several lessons. Later you will be making a rock collection. So, start looking for interesting rocks. Save them for your collection.

? Where are you likely to find rocks?

? How are igneous rocks formed?

? How are sedimentary rocks formed?

? How are metamorphic rocks formed?

Metamorphic

✏️ Characteristics of Rocks

Think about all of the different rocks you have seen.
Look at the pictures shown here.

⭕ Circle the words that could describe a rock.   ❌ Mark out the words that do not describe a rock.

Found in Earth's crust

Have babies

Hard

You can eat them

Squishy

Heavy

Smooth

Bumpy

Grow

Shiny

Different sizes

Colorful

Igneous Rocks

Have you ever seen pictures of a volcano? The fiery red liquid coming out of the volcano is lava.

Lava is melted rock that has come out of the earth. It is very hot inside the earth. The heat melts rocks that are inside the earth. When a volcano erupts, it shoots out some of that melted rock. Once the lava comes out of the earth, it cools down. When it becomes solid, it forms igneous rocks.

Igneous rocks can form inside the earth too. Remember that melted rock that is inside the earth is called magma. Sometimes magma moves to a cooler place inside the earth. When this happens, it hardens and forms igneous rocks.

? What is lava?

? What is one way that lava gets out of the earth's crust?

? What does lava turn into when it cools?

Name that Stone

There are many different kinds of igneous rocks. Granite is pretty when polished and used to make statues and counter tops. Pumice can be light enough to float in water because it has air holes all through it. It can be used to scrub off dead skin and make abrasives such as toothpaste.

There is another type of igneous rock that is black and shiny. It can be used to make arrowheads and other sharp tool blades.
Draw an arrowhead on the end of this arrow similar to the one on the previous page.
Color it black.

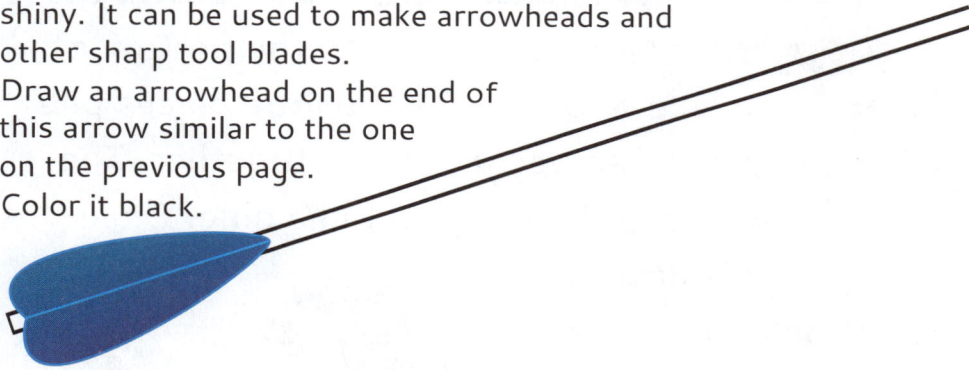

SEE OPTIONAL ACTIVITY

The name of this arrowhead's rock is in the puzzle. In the space below, write the letter of the alphabet that comes just BEFORE the one in the box. (For example, instead of T write an S.)

P	C	T	J	E	J	B	O

A B C D E F G H I J K L M N O P Q R S T U V W X Y Z

Sedimentary Rocks

Sedimentary rocks are usually made when little bits of rock, sand, dirt, or sea shells are washed away by moving water. These tiny pieces are called

sediment.

When the water slows down, these bits of rock settle to the bottom of a lake or ocean. As the little pieces pile up, they can become stuck together. The minerals in the water act like glue. Once the water dries, the sedimentary rock is left behind.

The Great Flood moved large amounts of dirt, sand, and rocks around. When the floodwaters dried up, they left behind most of the sedimentary rocks that we see today.

SEE OPTIONAL ACTIVITY

Playing with Sedimentary Rock

Some sedimentary rocks are not made by a flood. For example, clay bricks are made from bits of clay mixed with water. Concrete is made from rocks, sand, cement, and water. And sidewalk chalk is made from a mixture of gypsum and water. Think about different things that are made from bricks or cement. People use these sedimentary rocks to make walls, buildings, bird baths, stepping stones, and many other things. Use your sidewalk chalk to draw pictures of things made from sedimentary rock.

Scripture Trace

Never again shall there be a flood to destroy the earth.
Genesis 9:11b

Mixing Concrete

Concrete is made by mixing cement, water, sand, and rocks.
Look at the pictures and write the words in the blanks.

CEMENT

+

_____

+

_____

+

_____

= CONCRETE

Circle the things that are made of concrete.

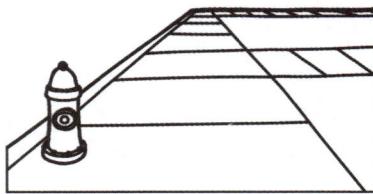

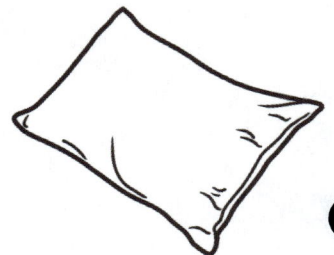

Fossils

Fossils are parts of dead plants and animals that have turned to rock. You might have seen dinosaur fossils at a museum. But only a very small number of fossils are from dinosaurs. Almost all fossils are sea creatures. They are mostly clams, snails, and other shellfish.

An animal can become a fossil if it is covered with mud very soon after it dies.

Only a few animals ever become fossils. Most animals that became fossils were covered with mud during Noah's Flood.

A fossilized shark's tooth

The floodwaters moved tons of mud and dirt around. This mud covered millions of sea creatures. These creatures eventually became fossils. Fossils are only found in sedimentary rocks.

Making a Fossil

Use a small plastic animal or sea shell to make a fossil. Pretend that it has died and is lying on the ground. Use play dough or modeling clay as your mud. Quickly cover the animal with the clay. Press the clay down. Then carefully remove the clay. Look at the shape of the clay. You have just made a mold for a fossil. The mud keeps this shape as the animal rots away. This shape is then filled with liquid. When the liquid hardens it becomes a fossil.

Mix a small amount of plaster of Paris with a small amount of water. Pour the liquid into the play dough or modeling clay mold you created. Allow it to dry overnight. Remove the fossil from the mold.

? What is a fossil?

? Do all animals become fossils when they die?

? What kind of rock would you look for if you wanted to find a fossil?

Fossil Matching

Match the fossil with the living creature.
Trace the names.

dragonfly

sea turtle

nautilus

crinoid

Fossil Fuels

Coal is an important type of sedimentary rock. It is burned for fuel in many power plants. It can be used to heat homes and other buildings. Some plants that were buried during the Great Flood turned into fossils, but many more plants turned into coal. Coal is made when plants are buried and then experience great heat and pressure.

Oil is another important source of fuel. Oil is made when millions of tiny sea creatures are buried and then experience heat and pressure. Oil is usually found in sedimentary rocks that contain fossils of sea creatures. Oil is used to make gasoline for cars and trucks.

Oil is also used to make plastic that is then made into many different items such as toys, cups, or trash cans.

Because coal and oil used to be plants and animals and are found in the same rocks as fossils, they are called

fossil fuels.

A third kind of fossil fuel is called

natural gas.

Natural gas is usually found near oil. Natural gas is used in stoves and is used to heat houses. These fossil fuels provide energy for our homes and cars.

? Name three types of fossil fuel.

? Why are these things called fossil fuels?

? What is coal made from?

? What is oil made from?

Match the Fuel

Fill in the missing vowels on the words below and match the word with the correct picture.

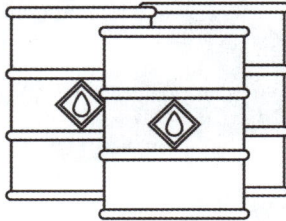

C _ _ L

G _ S

_ _ _ L

Metamorphic Rocks

So far, we have learned about two different kinds of rocks. Igneous rocks are made when melted rock cools. Sedimentary rocks are made when bits of dirt, sand, and rock are glued together. Now we are going to learn about a third kind of rock. This kind of rock is called metamorphic rock. Metamorphic simply means changed. When a caterpillar changes into a butterfly, we call that metamorphosis — something changing from one form into another.

Metamorphic rocks are rocks that have been changed from one kind into another kind. The rock starts out as either an igneous rock or a sedimentary rock. Heat is needed to make these rocks change. Pressure (squeezing) is also needed. When a rock experiences heat and pressure long enough, it changes into a new kind of rock.

These changed rocks are usually very hard.

Marble is one kind of metamorphic rock that is very hard. It is beautiful and can be polished to make it shiny. Marble is often used to make sculptures and floors.

? What does metamorphic mean?

? What two things are needed for a rock to change into a metamorphic rock?

? Name one kind of metamorphic rock.

? Have you ever seen a marble statue or floor?

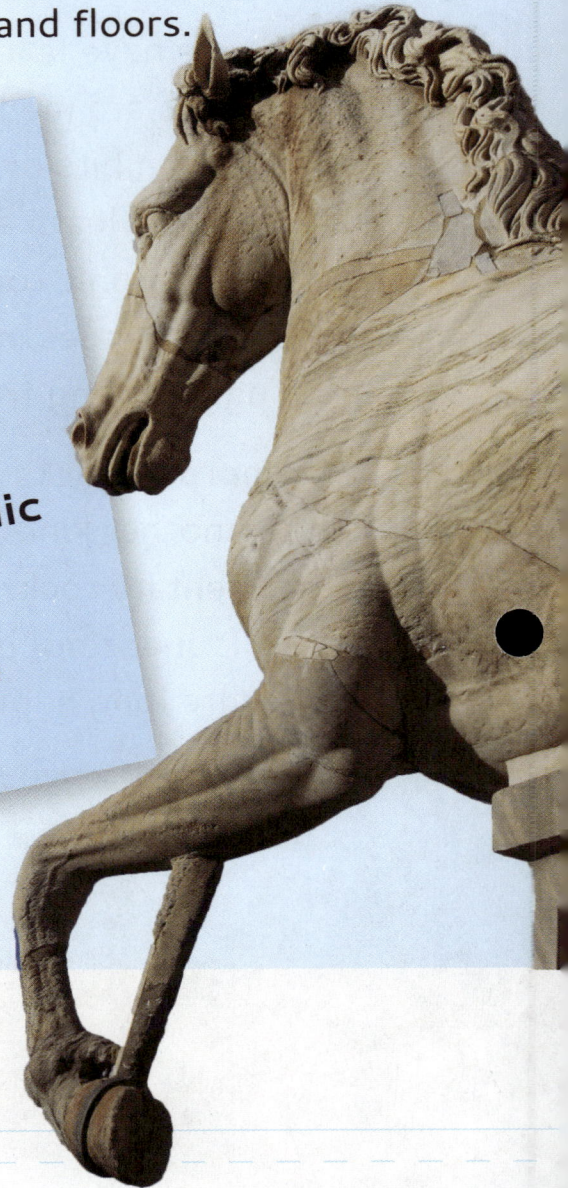

Scripture Trace

Therefore, if anyone is in Christ, he is a new creation.

2 Corinthians 5:17

Rock Review

Match the rock shapes at the bottom of the page and write the correct word part on each blank. Say each word part as you write it.

1. Rock made from cemented dirt, sand, and rocks is:

_____ _____ _____ _____ _____

2. Rock from lava or magma is:

_____ _____ _____

3. Rock that was changed by heat and pressure is:

_____ _____ _____ _____

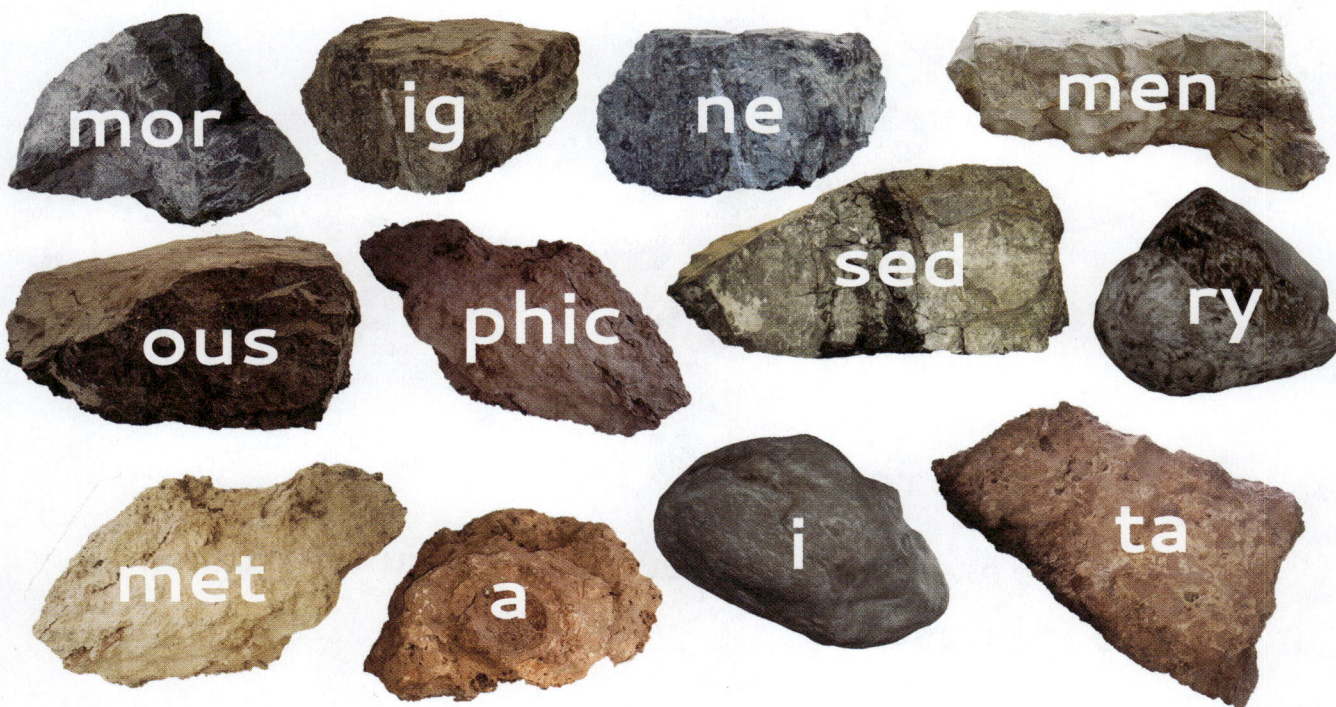

mor   ig   ne   men

ous   phic   sed   ry

met   a   i   ta

Minerals

Mica

Amethyst quartz

Rocks are made of

minerals. There are lots of minerals around your house.

You use minerals every day. Salt is a mineral that you probably eat every day. You will find other minerals in your food. You might find calcium, zinc, and iron in your breakfast cereal.

Other minerals are metals. Gold, silver, and copper are minerals, too. Minerals combine in many different ways to make all of the different rocks around us.

Copper

Gold

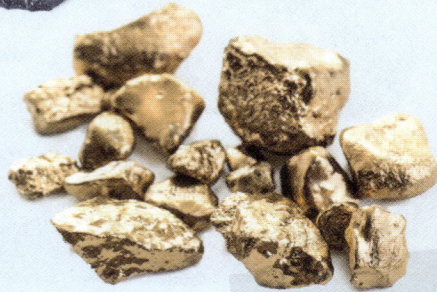

? What are rocks made from?

? What are some minerals you found around your house?

Mineral Scavenger Hunt

There are minerals all around you. Hunt around your house for the minerals shown here. Check off all of the minerals that you find. Hint! The pictures are clues for where you may find them.

☐ Salt    ☐ Diamond

☐ Calcium    ☐ Iron

☐ Aluminum    ☐ Zinc

☐ Copper    ☐ Gypsum

☐ Gold    ☐ Quartz

☐ Silver    ☐ Silica

Identifying Rocks

In lesson 9, you started collecting rocks. Look at some of the rocks you have so far. You can learn to identify each rock. To *identify* means to tell what kind it is. First look at the color of the rock. Different kinds of rocks are different colors. Next, look at how shiny the rock is. Is it sparkly, shiny, or dull? Also, look for **crystals**. Many rocks do not have any crystals. Some rocks have very small crystals. Others have large crystals. Crystals can help tell how the rock was formed. Finally, get a rocks and minerals guidebook. Compare your rock to the pictures in the book. This should help you identify your rock.

? What are three things you can look at to help you sort rocks?

? What does your favorite rock look like?

Collecting Rocks

Keep collecting rocks. Practice sorting your rocks by color, shine, and crystals just like you will do on this worksheet.

Sorting Rocks

Look at each row of rocks. Three of them are alike in some way.
Decide how they are alike. ✕ Mark out the one that is different.

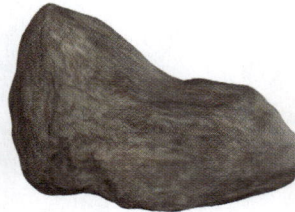

Rock Cycle

One type of rock can change into another type of rock. This process is called the

rock cycle.

Melted rock

Heat can melt a <u>sedimentary</u> rock. When it cools, it becomes an <u>igneous</u> rock.

A <u>metamorphic</u> rock can be melted. When it cools, it becomes an <u>igneous</u> rock.

? How can an igneous rock become a metamorphic rock?

? How can a metamorphic rock become a sedimentary rock?

? How can a sedimentary rock become an igneous rock?

Igneous rock

An igneous rock can also become a sedimentary rock. Over time, small pieces of the igneous rock can be broken off by wind and rain.

An igneous rock can change into a metamorphic rock. It takes heat and pressure to do this.

Sediments

These pieces could be glued together to form a sedimentary rock.

Pieces can be broken off of a metamorphic rock. When they are glued together, they become part of a sedimentary rock.

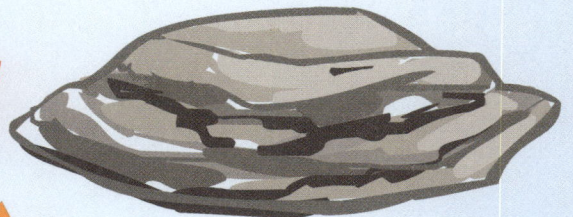

Sedimentary rock

If there is great heat and pressure a sedimentary rock might become a metamorphic rock instead.

Metamorphic rock

Color the Arrows

Color the arrows that show rock melting red.

Color the arrows that show rock cooling blue.

Color the arrows that show rocks being broken into small pieces brown.

Color arrows leading to metamorphic rocks orange.

Color arrows leading to sedimentary rock gray.

Melted rock

Igneous rock

Metamorphic rock

Sediments

Sedimentary rock

Gems

Some minerals are very special. Have you ever seen a diamond ring sparkling in the light? How about a red ruby or a green emerald? Stones that can be cut to reflect light are called

_____

gems. Other popular gems include sapphire, topaz, and amethyst. People wear gems as jewelry. Gems were important to people in the Bible, too. The high priest in the Bible wore a special breastplate that contained 12 different gems (Exodus 28:15–21).

? What is a gem?

? Have you ever seen a diamond?

? Why do you think people use gems in jewelry?

? Ask your parents to show you any gems that they own.

Dot to Dot

Connect the dots 1–20 to see a special gem.

name _____

4

5

1

3   6

8

2

7

20   19   10   9

18   11

15   14

17   12

16   13

Scripture Trace

And he took the crown of their king from his head. In it was a precious stone, and it was placed on David's head.

2 Samuel 12:30

Breastplate Coloring Sheet

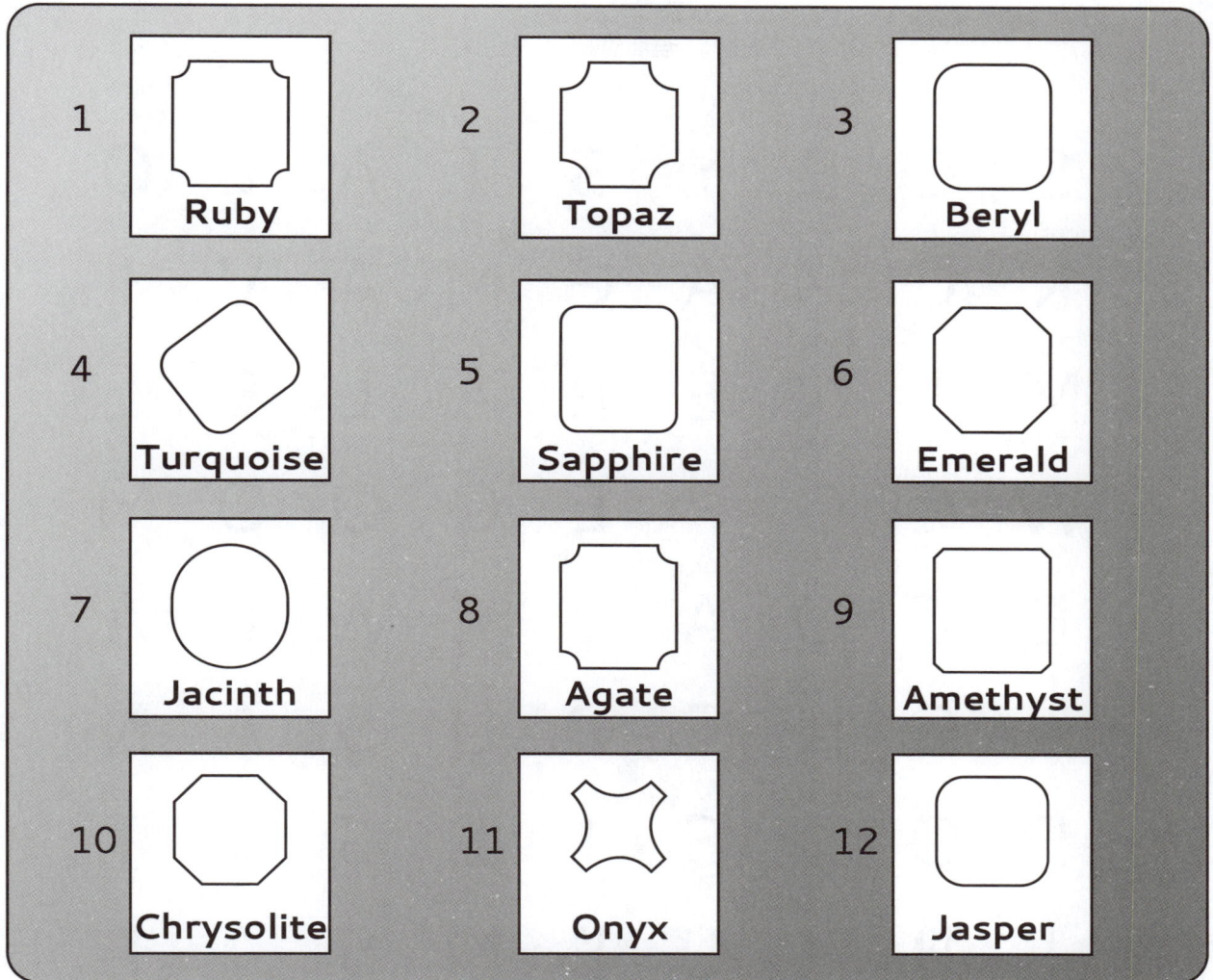

1 Ruby	2 Topaz	3 Beryl
4 Turquoise	5 Sapphire	6 Emerald
7 Jacinth	8 Agate	9 Amethyst
10 Chrysolite	11 Onyx	12 Jasper

Color each gem a different color (see below for suggestions).

Ruby: Dark red

Turquoise: Light greenish blue

Jacinth: Orange

Chrysolite: Green

Topaz: Yellow to brown

Sapphire: Dark blue

Agate: Striped or marbled in any color

Onyx: Grayish black

Beryl: Green, bluish green, yellow, pink, or white

Emerald: Bright green

Amethyst: Dark purple

Jasper: Dark green

How many rows are there?

How many gems are there in each row?

Unit Vocabulary Review

```
S C R Y S T A L O
K C M A G M A V I
M I N E R A L S L
W C O R E Q O O M
E C O A L C W L A
I G N E O U S A N
F O S S I L S V T
I P X G E M S A L
E R O C R U S T E
```

COAL	FOSSILS	LAVA
CORE	GEMS	MAGMA
CRUST	IGNEOUS	MANTLE
CRYSTAL	OIL	MINERALS

Mountains and Movement

Planet Earth
for Beginners

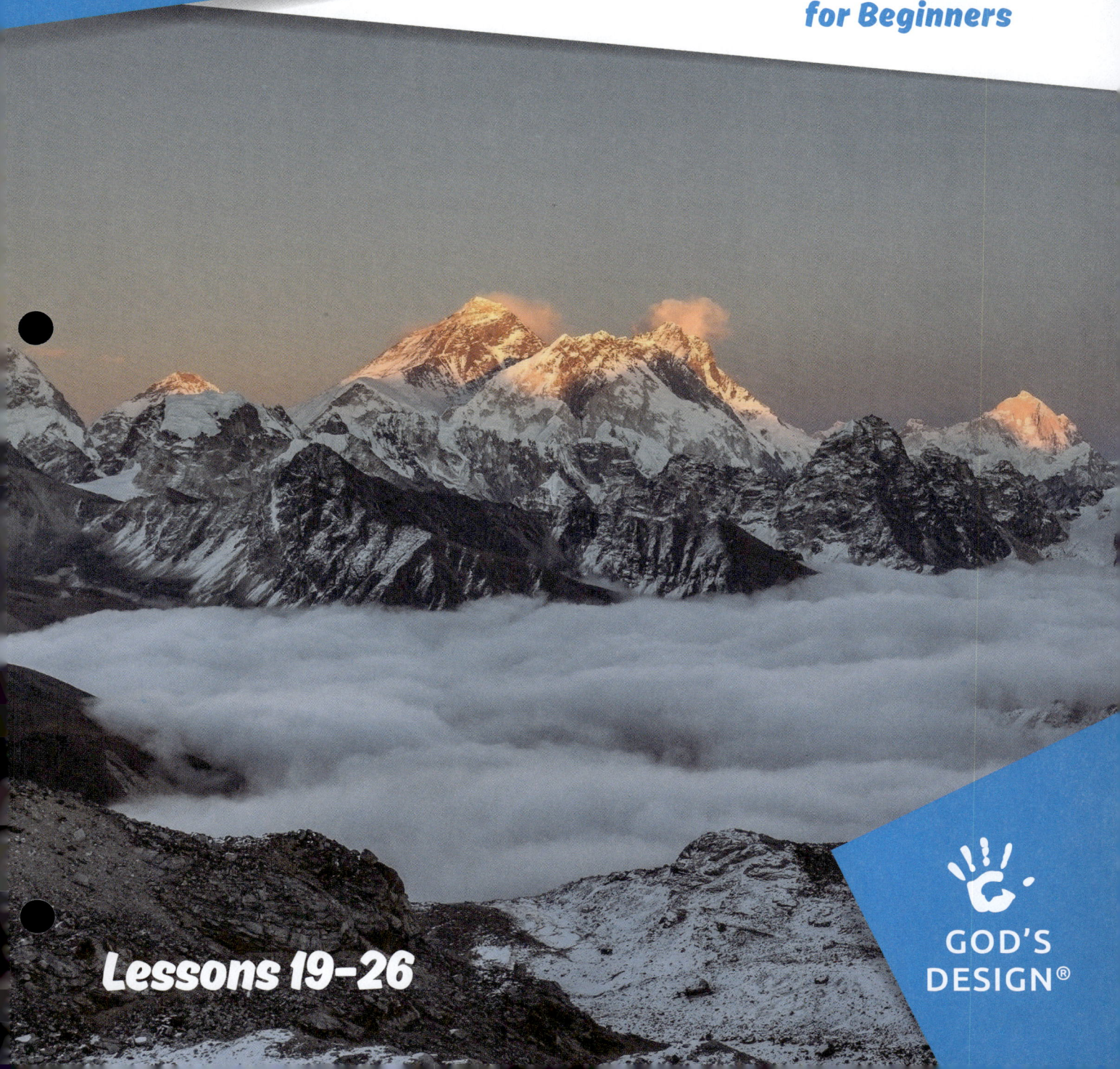

GOD'S DESIGN®

Lessons 19-26

The Earth Has Plates

The earth has areas of land and areas of water. The large pieces of land are called _continents_. The large areas of water are called _oceans_. Today there are seven continents: North America, South America, Europe, Asia, Africa, Australia, and Antarctica. Look at a globe to see where each continent is. There are five oceans: Pacific, Atlantic, Indian, Arctic, and Southern. Again, see if you can find each ocean on a globe.

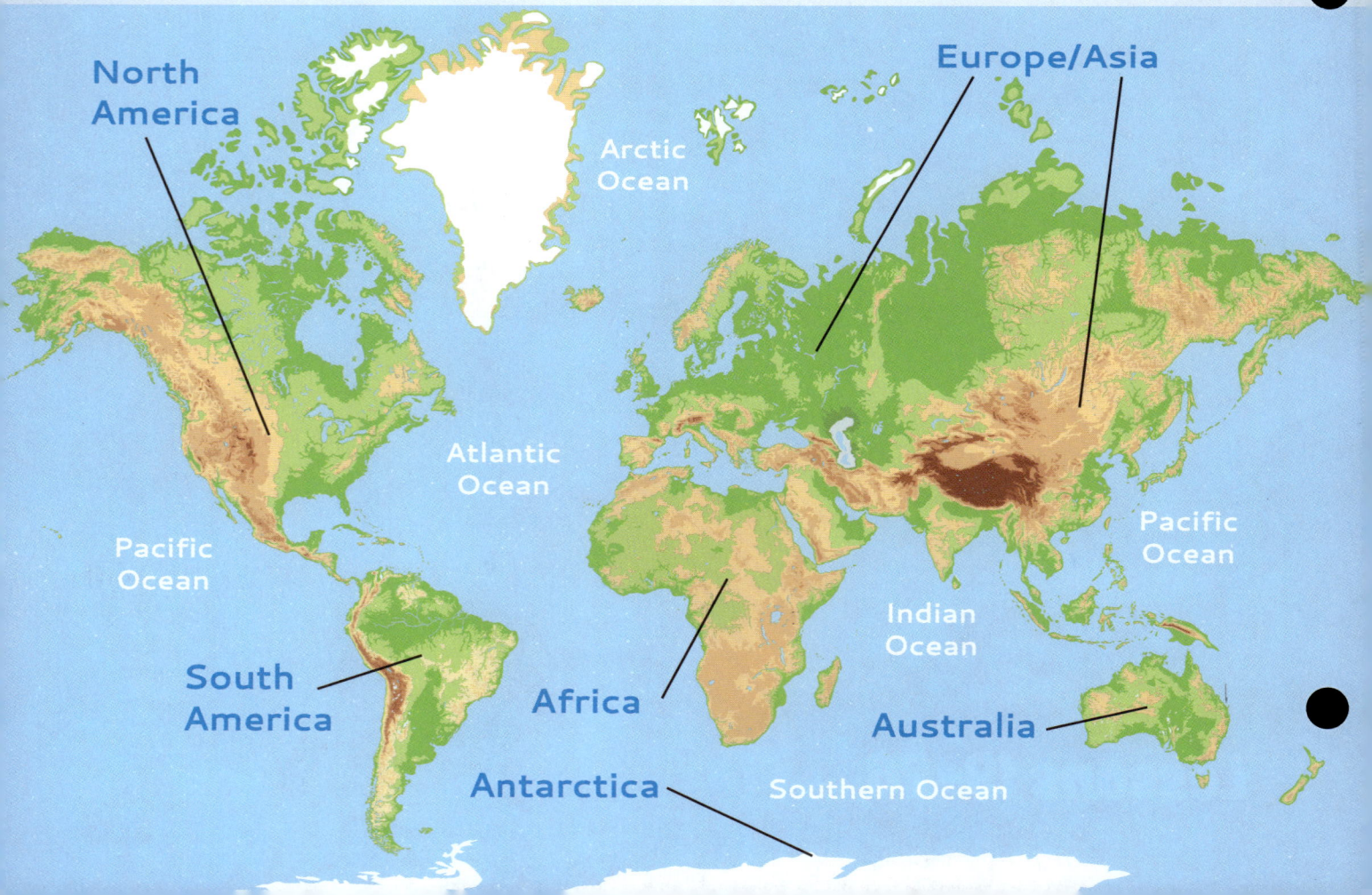

North America

Europe/Asia

Arctic Ocean

Atlantic Ocean

Pacific Ocean

Pacific Ocean

Indian Ocean

South America

Africa

Australia

Antarctica

Southern Ocean

Moving Plates

The earth's plates moved quickly during the Flood. But today the plates move slowly. You can see how this works by doing the following activity. Spread a thick layer of frosting on one half of a baking sheet. This represents the earth's mantle. Place two graham cracker squares on top of the frosting. This represents two plates in the earth's crust that rest on the mantle. Try sliding the plates toward each other. You should see that frosting was pushed up. What happened when the plates met? Did they begin to crack? Did one slide under the other? Sometimes one of the earth's plates will slide under another.

Place two crackers where they are touching each other. Now try pushing one plate away from you and sliding the other toward you so their edges rub together. What happened to the crackers? Small crumbs will form. When the earth's plates slide against each other, they sometimes slip. This causes an earthquake.

Finally, push the plates away from each other. You might see frosting filling in some of the gap. When earth's plates move away from each other, magma fills in the space and hardens into new rock. This happens most often on the sea floor.

Before the Flood, the earth looked different than it does today. People who study the earth believe that at one time there was only one large continent. This continent has been given the name

Rodinia.

During the Genesis Flood the crust of the earth cracked. This allowed water that was trapped below ground to shoot up. This was where most of the water came from for the Flood. These cracks in the earth's crust allowed the land to break up.

The different pieces of the crust are called

<u>plates</u>. These plates collided (smashed into each other) to form moutain ranges and moved apart to form the continents and ocean basins that we have today.

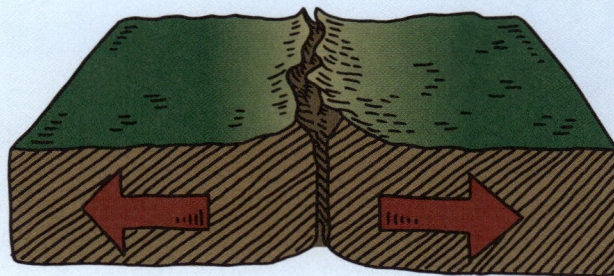

? What are large areas of land called?

? What are large areas of water called?

? What continent do you live on?

? How many continents do people believe there were before the Flood?

? How many continents are there today? Memorize the names of the continents.

Continents Before and After

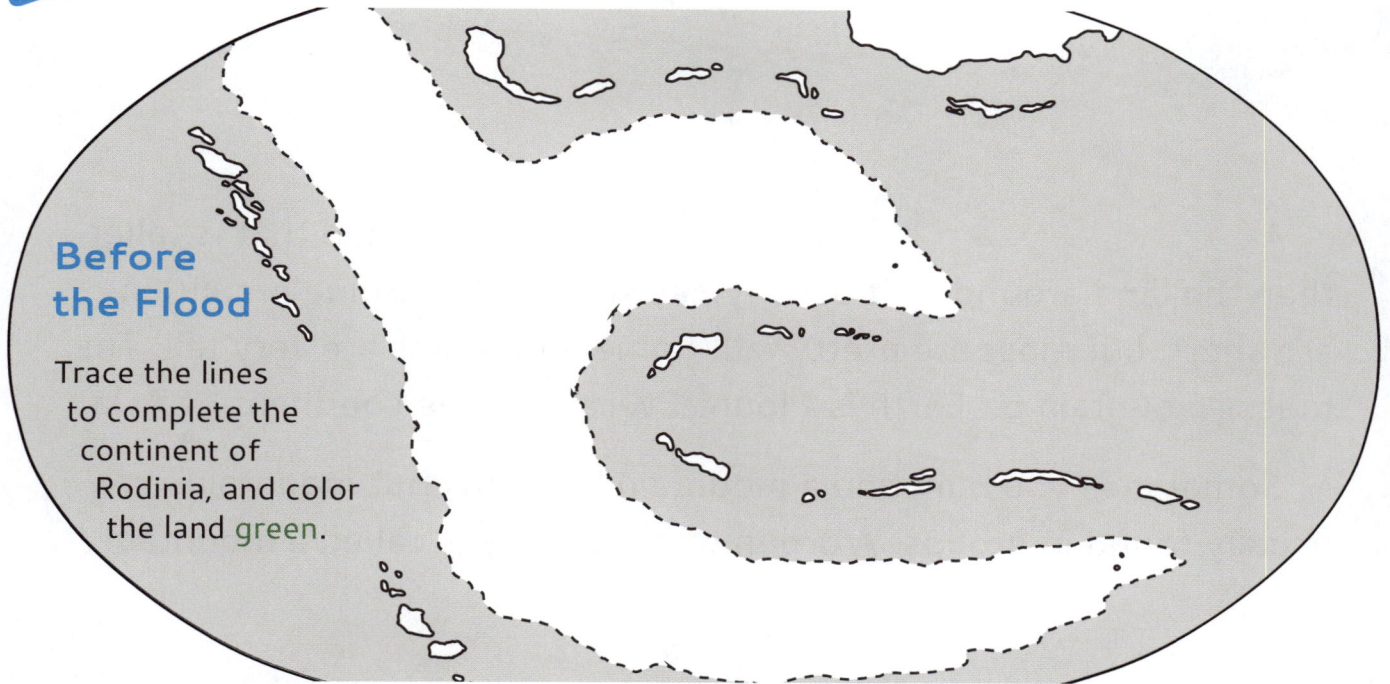

Before the Flood

Trace the lines to complete the continent of Rodinia, and color the land green.

Write the number of each continent name in the correct space on the map. Then use the code to color each continent.

1—North America	5—Africa	1—Blue	5—Orange
2—South America	6—Australia	2—Red	6—Brown
3—Europe	7—Antarctica	3—Green	7—Purple
4—Asia		4—Yellow	

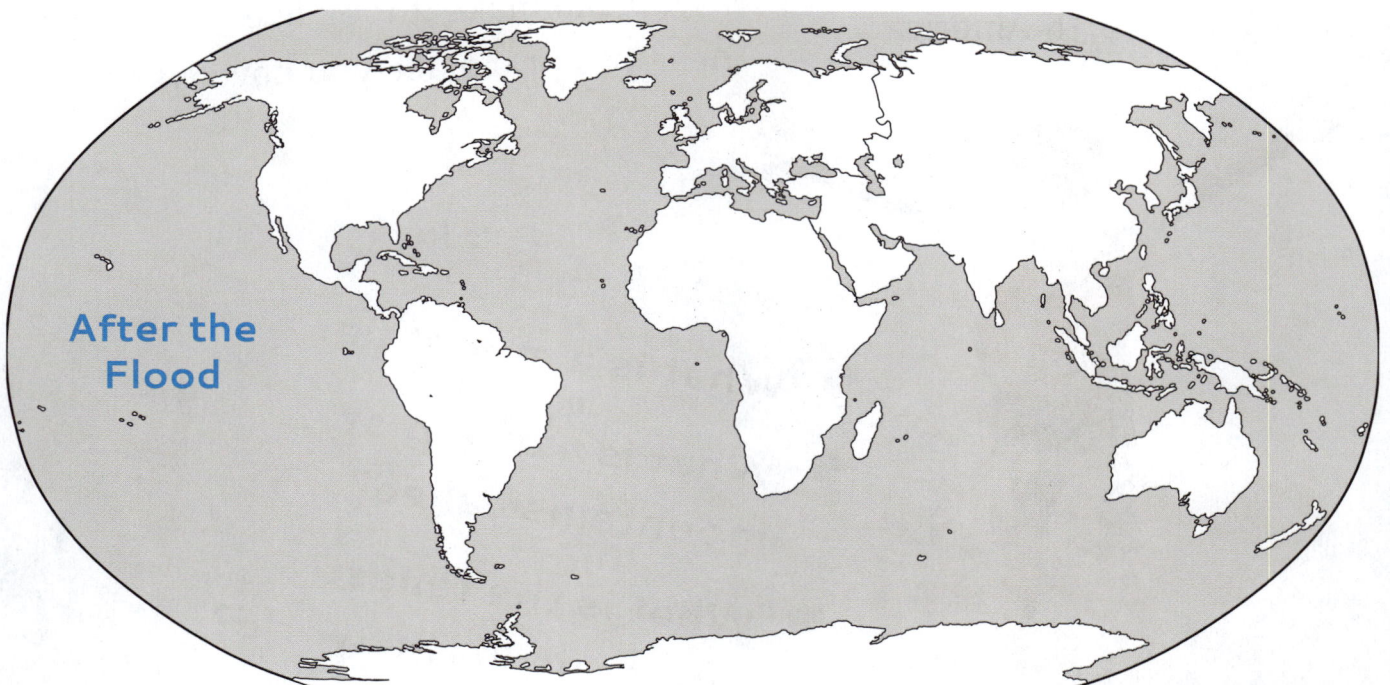

After the Flood

Mountains

A __mountain__ is an area of land that is taller than the land around it. It usually has steep sides. Some mountains are short, but most are pretty tall. Some mountains are very tall. The tallest mountain on Earth is Mount Everest on the continent of Asia.

Sometimes you may find a mountain by itself. But mountains are usually found in groups. A group of mountains is called a mountain

__range__. The Cascades, Sierra Nevadas, Rocky Mountains, Ozarks, and Appalachian Mountains are some important mountain ranges in North America.

Build a Mountain Range

Use modeling clay or play dough to build your own mountain range. See if you can make it look like mountains near you or a place you have visited.

? What is a mountain?

? What is a group of mountains called?

? What is the tallest mountain in the world?

What Lives in the Mountains

Trace the gray lines to complete the mountain range. Color the rocks brown. Count the trees and the animals.

How many animals can you count?

Types of Mountains

Mountains can be formed in many different ways. Some mountains were made by volcanoes. Volcanoes shoot out lava, rocks, and ash. If the volcano erupts a lot, these materials pile up. They can form a new mountain.

Sand dunes are mountains of sand. Sand dunes are formed when wind blows sand into giant piles. Sometimes the sand is trapped in an area and can form very tall sand dunes.

Other mountains were pushed up. The plates in the earth's crust push against each other. When a plate is pushed on from opposite sides, the land in the middle can be forced up. This can form new mountains.

? How do volcanoes form new mountains?

? What are sand dunes?

? How were some sand dunes formed?

? What happened when the earth's plates pushed against each other?

✏️ **Scripture Trace**

For behold, he who forms the mountains and creates the wind,

Amos 4:13a

Mountains Forming

Look at the pictures. Circle  the letter for the type of mountain it shows.
V = volcano, S = sand dune, P = pushed up.

1

V   S   P

2

V   S   P

3

V   S   P

Making Mountains

Lay several sheets of newspaper or paper towels on a table. Sprinkle a little water on the paper. Do not make the paper soggy. The paper represents one of the plates in the earth's crust. Place one hand on each side of the paper. Slowly push the edges of the paper toward the middle. This should cause the paper to push up in the middle. These are mountains that are formed when the plate is pushed on from two sides. Let the paper dry overnight. The mountains will keep their shape.

Earthquakes

An __earthquake__ is when the ground suddenly shakes.

Some of the earth's plates are pushing against each other. When one plate suddenly slips, it causes an earthquake. Earthquakes happen several times a day somewhere on earth. But most earthquakes are too small for people to feel. Strong earthquakes happen a few times each year.

A strong earthquake may be followed by several smaller earthquakes. These small earthquakes are called __aftershocks__.

An earthquake can also cause huge waves called

tsunamis.

These giant waves can crash into land that is very far away from where the earthquake happened.

? **What is an earthquake?**

? **What is an aftershock?**

? **What is a tsunami?**

? **Have you ever felt an earthquake?**

All Fall Down

Earthquakes can cause buildings to fall down. See if you can build a building that can survive an earthquake. Use blocks to make a building on a table. Gently shake the table. Does your building fall down? How much shaking does it take before the building falls?

Scripture Trace

There was a great earthquake, so that the foundations of the prison were shaken.

Acts 16:26

Recording Earthquakes

Scientists use a machine called a seismograph to record how strong an earthquake is. The larger the bumps on the paper, the stronger the earthquake.

Circle ◯ the part that shows a large earthquake. Draw a square ▢ around the part that shows a small earthquake, or aftershock. Place a checkmark ✓ on the place that shows no earthquake.

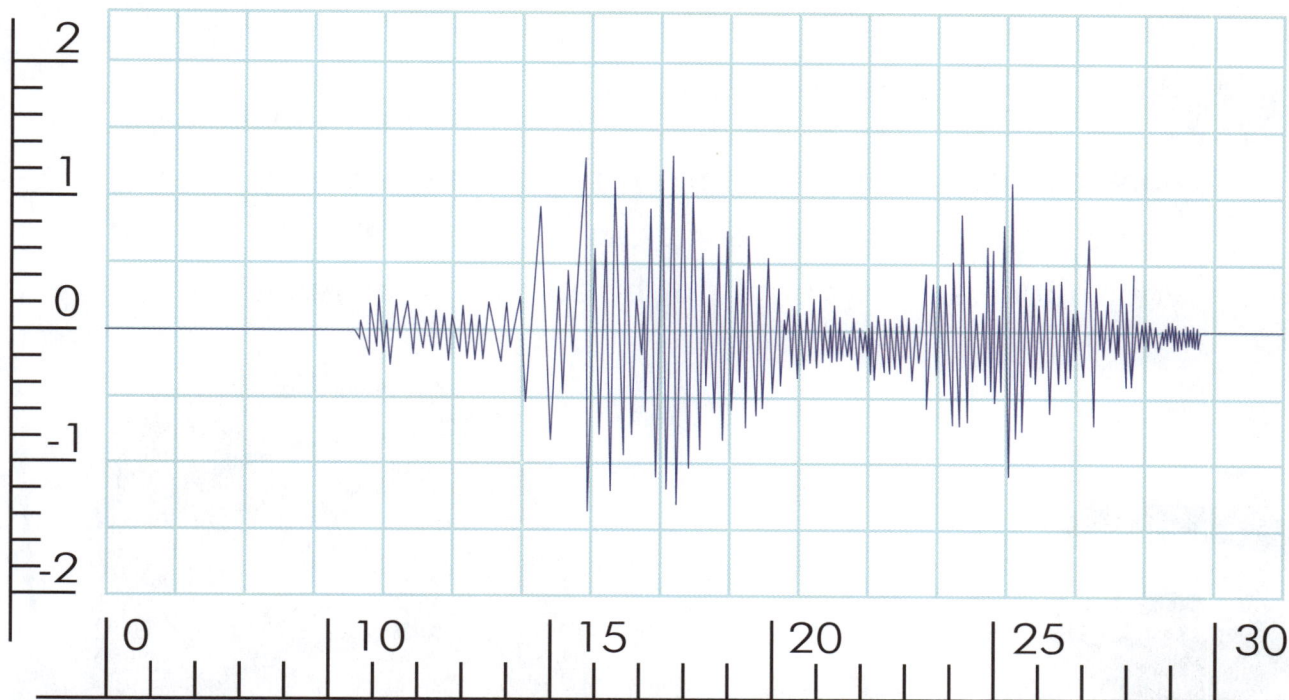

Draw lines below to show an earthquake followed by two aftershocks. Remember that aftershocks are not as strong as the original earthquake.

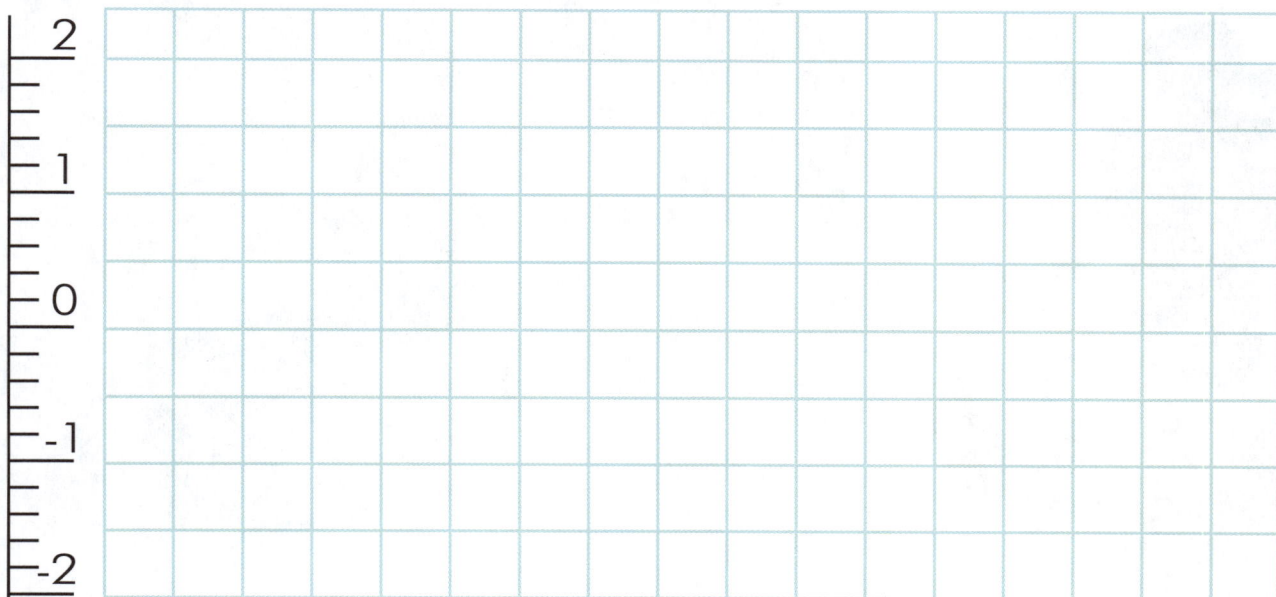

Preparing for an Emergency

An earthquake can be a serious

emergency. An earthquake can damage your house. You might not have electricity for a while after an earthquake. It can damage the roads around your house so you might not be able to go anywhere.

There are other types of emergencies too. You might get snowed in and not be able to leave your house. Your house might catch on fire. Maybe the roads are closed because of a bad storm. You might not be able to get to a store.

You and your family need to be prepared for an emergency. Have a family meeting. Find out what you should do in case of each type of emergency. Practice doing those things.

? **What are some things that can cause an emergency?**

? **How can you prepare for an emergency?**

Make an Emergency Kit

Part of being prepared for an emergency is having an emergency kit. Help your family gather the items for an emergency kit. Your kit should include:

☐ Non-perishable food (such as dried fruit, canned food, or peanut butter)

☐ Can opener (manual)

☐ First aid kit

☐ Flashlight with extra batteries

☐ Matches

☐ Toothbrush, toothpaste, soap

☐ Paper plates, plastic cups and utensils, paper towels

☐ Water (at least a gallon per person, per day)

☐ Sleeping bag or warm blanket for each person

In Case of Emergency

✗ Mark out the picture showing what you should not do in case of the pictured emergency. Then color the picture of what you should do.

House on fire ———————————————

Power is out ———————————————

Bad storm with lightning ———————————————

318   *Heaven and Earth for Beginners*

Volcanoes

It is hot inside the earth. This heat melts rocks. Sometimes this melted rock builds up pressure. When there is enough pressure, this can make a volcano erupt. Most volcanoes shoot out lava.

Lava is melted rock that has come out of the earth. Volcanoes also shoot out ash and cinders. __Ash__ is very small bits of rock. __Cinders__ are medium-sized bits of rock. You will also see __steam__ coming out of a volcano. All of these things can be very dangerous. You definitely don't want to be near a volcano when it is erupting.

Some volcanoes are very **active**. They rumble and smoke. They erupt all the time. Other volcanoes have erupted then stopped. They are not erupting right now. But they could erupt again at any time.

Other volcanoes are called **dormant** volcanoes. This means they are asleep. They have not erupted in the past 50 years. But it is possible that they could erupt in the future.

Finally, some volcanoes are called **extinct** volcanoes. These are volcanoes that people think will never erupt again.

? What are some things that come out of volcanoes?

? If a volcano has not erupted for a long time but could erupt again is it dormant or extinct?

Volcano Maze

name _____

Collect the letters that help you get through the maze and write them in the blanks below to reveal four things that come out of a volcano.

Hint!
All four
words are
found in this lesson.
The number of letters
in each word are given below.

_____   _____   _____   _____

7 LETTERS   3 LETTERS   4 LETTERS   5 LETTERS

Volcano Types

Volcanoes form mountains as they send out lava, ash, and cinders. The mountains are different shapes depending on what is coming out of the volcano. If a volcano shoots out mostly lava, it forms a rounded mountain. These mountains look like shields, so they are called

__shield__ volcanoes.

If a volcano shoots out mostly ash and cinders, it forms a mountain that has steep sides. It is shaped like an upside-down cone. These are

called __cinder__

__cone__ volcanoes. Cinder cone volcanoes are usually not very tall.

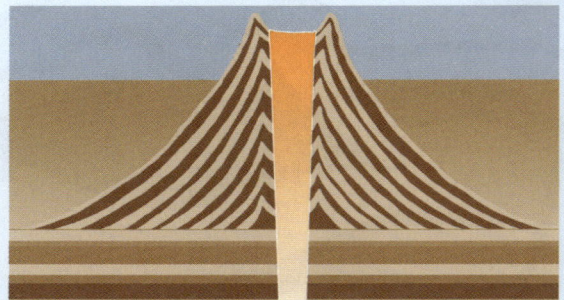

Mauna Loa, a shield volcano in Hawaii, is the largest volcano in the world.

Most volcanoes shoot out lava for a while. Then they shoot out ash and cinders for a while. Then they shoot out more lava. These are called

composite volcanoes. They are also cone-shaped but are much taller than cinder cones. Mount Fuji and Mount St. Helens are two famous composite volcanoes.

? Why do different volcanoes have different shapes?

? Which shape is most common?

Cool Volcano

Place ice cream in a dish. Shape it to look like a volcano. Scoop out a crater on the top. Fill the crater with chocolate syrup lava. Let some of the lava flow down the sides. Sprinkle cookie crumb ashes on top of your volcano. Now enjoy a cool treat.

If you would rather have your volcano for dinner, you can make the volcano out of mashed potatoes, use gravy for the lava, and sprinkle it with bread crumb ashes.

Capulin in New Mexico is an extinct cinder cone volcano.

Mt. Fuji in Japan is an extinct composite volcano.

Volcano Match Up

Follow the lines to discover which type of volcano is pictured.

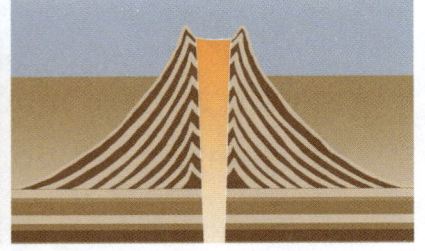

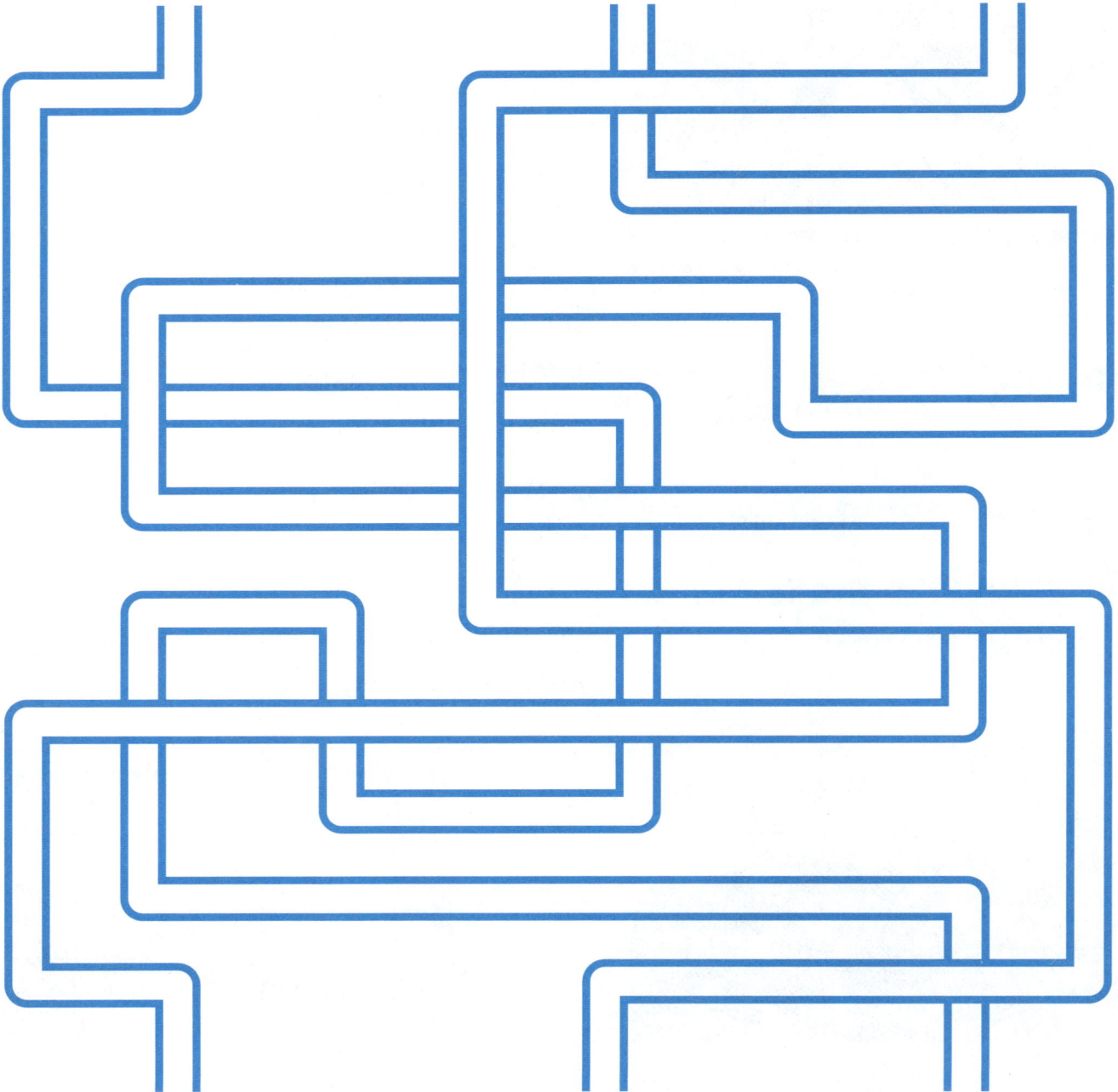

COMPOSITE   **CINDER CONE**   **SHIELD**

Mount St. Helens

One of the most famous volcanoes in the United States is Mount Saint Helens. This volcano is located in southwest Washington State. Mount St. Helens had been quiet for 123 years. But on May 18, 1980, it erupted with a huge explosion. The explosion blew 1,300 feet (400 m) off the top of the mountain. The explosion was heard 200 miles (320 km) away. This was an awesome and frightening eruption. The volcano began erupting again in 2004 and continued until 2008. But the new eruption was not nearly as big as the first one. Now Mount St. Helens is quiet. But it could erupt again in the future.

The 1980 eruption burned and blew down 230 square miles of the surrounding forest.

Scientists were able to learn many things from studying the Mount St. Helens eruptions. They saw 25 feet of volcanic ash build up in only one day. This ash settled in many different layers and looked just like the sedimentary rock in many parts of the earth. This shows that sedimentary rock can form very quickly.

Another interesting thing was that a huge amount of water and mud came out of the volcano. This water and mud dug a 100-foot-deep

Volcano Song

Sing the following words to the tune of "All Around the Mulberry Bush," also known as "Pop Goes the Weasel." When you get to the last line, imitate a volcano that is erupting.

The rocks inside the earth get hot.

They melt and become lava.

The pressure builds and builds until

POP goes the volcano!

canyon or valley in a very short time. So, we know that large canyons can be formed quickly if there is a lot of water. These things agree with what the Bible says and help to show that God's Word is true. The earth is young. And many of the things we see around us show that there was a great flood.

? **Is Mount St. Helens an active, dormant, or extinct volcano?**

? **What happened when it erupted in 1980?**

Scripture Trace

Every word of God proves true; he is a shield to those who take refuge in him.

Proverbs 30:5

Unit Vocabulary Review

Fill in the missing letters in the words below. Then unscramble the letters you added to answer the question below.

Contin___nts

Oc___ans

___odinia

___ ___ountain Ran___e

Sa___d Dune

___arthquake

Aftersho___k

You should always try to prepare for an _____y.

Answer the following questions about volcanoes using vocabulary words you learned in the lessons.

What are three things that often come out of volcanoes?

A _____

C _____

S _____

What are three types of volcanoes?

S _____

C _____

Com _____

If a volcano has erupted recently it is a _____.

If a volcano has not erupted for at least 50 years it is d _____.

If a volcano is not expected to ever erupt again it is e _____.

Water And Erosion

UNIT
4

Planet Earth
for Beginners

Lessons 27-35

GOD'S
DESIGN®

Geysers

Have you ever heard of Old Faithful? It is a geyser in Yellowstone National Park. A

geyser is

an opening in the ground where hot water and steam shoot out. Old Faithful erupts about once every 90 minutes.

Make a Geyser

This activity should be done outside. Fill a straw with water. Put your finger over the end of the straw to keep the water inside. Tip your head back. Place the straw in your mouth. Blow the water out of the straw. The water will shoot up into the air. You have made a geyser.

A geyser forms when water inside the earth is heated up. Hot water takes up more room than cool water. This creates pressure under the ground because the hot water pushes out against the rocks. The water moves up until it finds an opening in the ground. If the pressure is great enough, the water will shoot out of the ground in a huge fountain. This is a geyser.

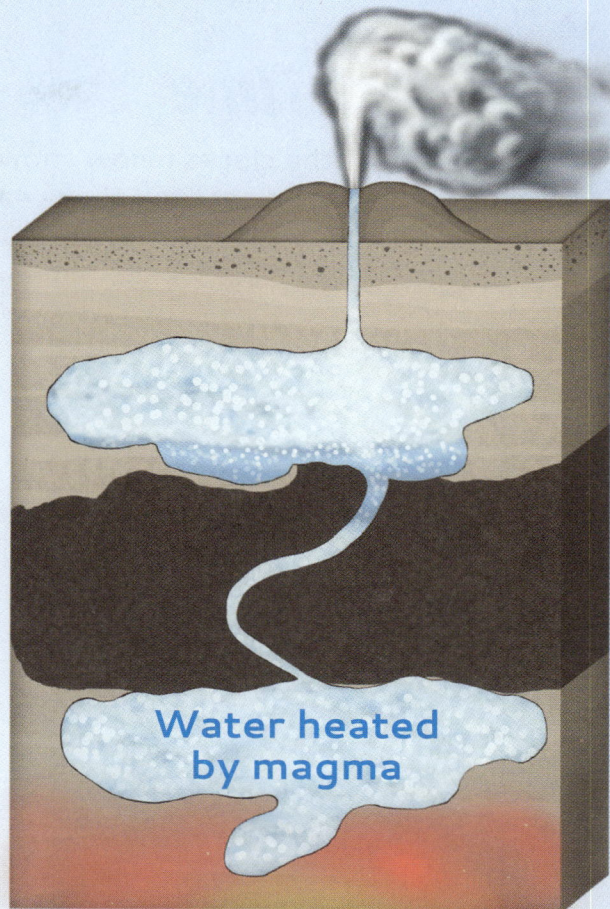

Water heated by magma

? Why does water shoot out of a geyser?

? Where should you go if you want to see Old Faithful?

✏️ **Scripture Trace**

On that day all the fountains of the great deep burst forth.

Genesis 7:11b

Geyser Time

Imagine that the clocks below show the times you can see Old Faithful erupt. Circle the right time above each clock.

1. 1:00 or 6:00

2. 12:00 or 2:30

3. 9:00 or 4:00

Old Faithful is about to erupt. Draw water and steam shooting out of the top of this famous geyser.

4. 3:00 or 5:30

5. 7:00 or 10:00

6. 11:30 or 8:30

Erosion

When rain falls it hits against rocks. This can break off small bits of the rock. The rocks are worn away a little at a time. This

is **erosion**. Sometimes water fills in small cracks in a rock. When the water freezes it expands. This means it takes up more space. The ice pushes on the rock and makes the crack bigger. Eventually, the rock will break apart. This is erosion, too.

Wind can also cause rocks to break into pieces. Wind blows bits of rock, glass, or other materials against rocks. This breaks off tiny pieces of rock. Strong winds can eventually wear away large amounts of rock.

Erosion is constantly happening around us. It usually happens slowly. Only tiny bits of rock are worn away at a time. But after many years, you can see a difference in the rock. Sometimes erosion happens very quickly. When there is a flood, the rushing water can wear away a huge amount of rock in a very short time.

SEE OPTIONAL ACTIVITY

? What is erosion?

? What are the two main things that wear away the rocks?

? Does erosion usually happen quickly or slowly?

? What can cause erosion to happen very quickly?

Wind and Water

Color the 🌀 wind-worn blocks **brown.**   Color the 💧 rain-worn blocks **blue.**

Write the revealed word in the space below.

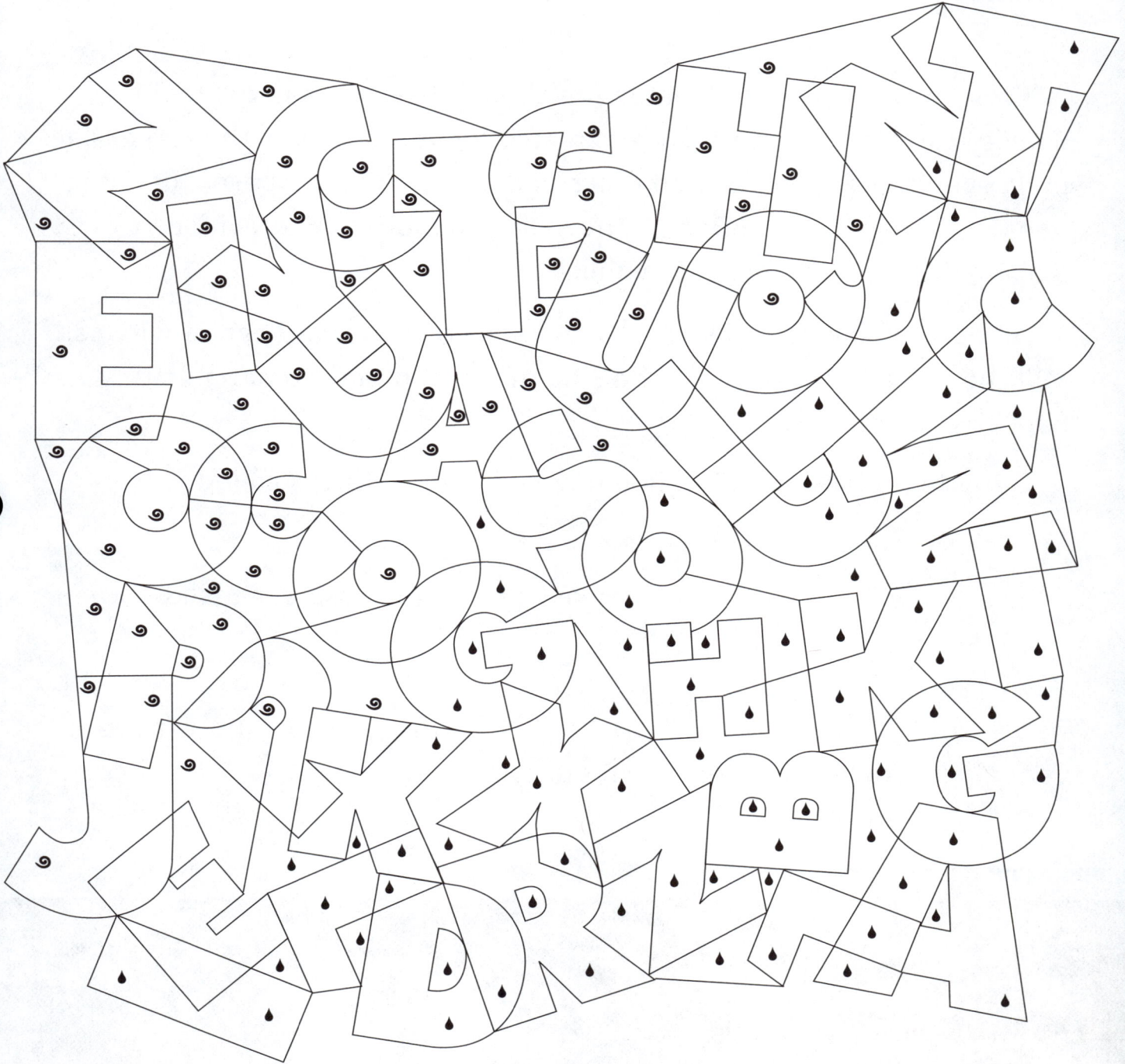

_____

- - - - - - - - - - - - - - -

_____

Landslides

Erosion from wind and water can change the way an area looks. Gravity can also change the way an area looks. Gravity is the force that pulls everything down. When rocks and dirt are on a flat area, gravity does not make it move. But on the side of a mountain, gravity can make rocks and dirt move downhill.

An earthquake can make rocks and soil come loose. Then gravity can pull the dirt and rocks down the side of a mountain. Sometimes this happens very slowly.

But when rocks and dirt slide down very quickly it is called a

landslide . Heavy rains can cause a landslide too.

An avalanche is when ice and snow move very quickly down the side of a mountain.

👆 Make a Landslide

Place an inch of dirt or potting soil in the bottom of a baking dish. Press it down firmly. Lift the end of the dish. Raise it slowly. Hold the edge of the pan several inches above the table. This makes a steep slope. Shake the pan gently. This is what happens during an earthquake. Did the soil move down the pan? If not, make the slope steeper and try again. An earthquake loosens the soil so that it can slide down a hill causing a landslide.

Set the pan down. Spread the soil back out and press it down again. Lift the edge of the pan again. This time pour some water on the soil. Did the soil slide down again? Rain can loosen the soil on a mountain and cause a landslide, too.

? What force pulls everything down?

? Is an avalanche more likely to happen in the summer or the winter?

✏️ Scripture Trace

A great and strong wind tore the mountains and broke in pieces the rocks before the LORD.   I Kings 19:11

Going Down the Hill

Find your way through the landslide to the road.
Collect the letters as you go to answer the question below.
Then color the picture.

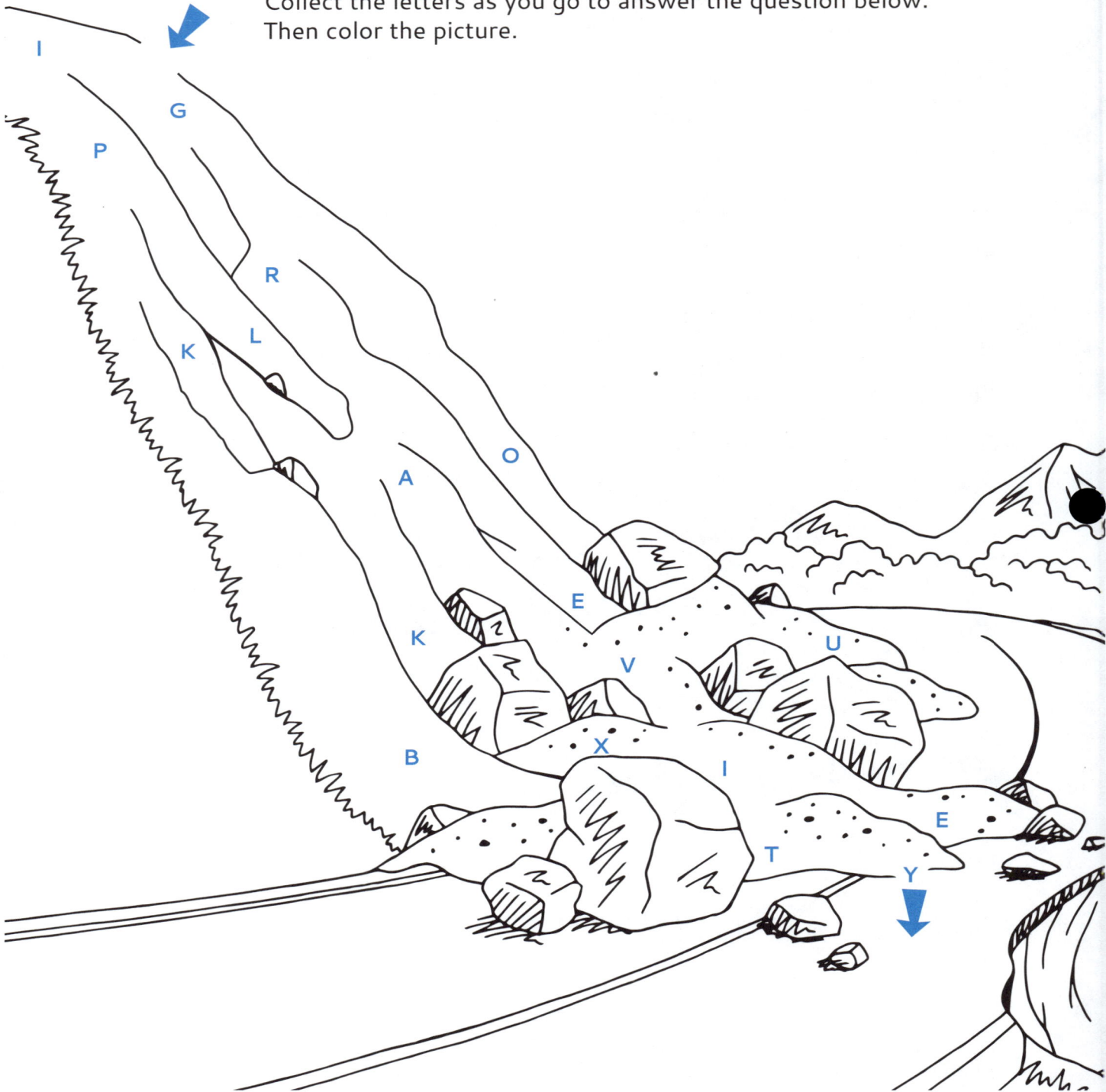

I
G
P
R
K L
A
O
E
K
V U
B
X
I
T
E
Y

What force pulls everything down? _____

_____

Stream Erosion

Rain and wind can wear away the surface of a rock. This is usually a slow process. But moving water can wear away rocks or soil very quickly. This is called _stream erosion_.

Gravity pulls water downhill. The steeper the hill, the faster the water will flow. The water that is flowing in rivers and streams breaks off bits of rock. The faster the water moves, the more rocks and dirt it can break off. The water can carry these small pieces down a mountain. It can carry them all the way to a lake or even to the ocean.

If a farmer wants to grow crops on the side of a hill, he must terrace the hill.

A _terrace_ is a flat area cut out of the side of the hill.

Terraced land looks like a set of steps cut into the side of a mountain. Water coming down the hill slows down when it reaches each step, so it does not carry away as much soil when it flows down the hill.

? How is stream erosion different from rain or wind erosion?

? Why does water flow downhill?

? Does water flow quickly or slowly down a steep hill?

Flowing Water

Water flows faster down a steeper slope. You can see this by doing the following experiment.

Place a baking sheet on a table. Drop a few drops of water at one end. Watch where the water goes. It will pretty much stay in one spot.

Dry the baking sheet. Place a small book under one end. Now drop a few drops of water on the raised end of the baking sheet. Where did the water go? It flowed down the hill. How fast did it flow down the hill?

Place another book under the baking sheet. Drop a few more drops at the top edge of the baking sheet. Did the water flow faster? It should. The steeper the hill, the faster the water will flow. When water flows quickly, it can carry away more dirt and rocks than when it flows slowly.

Scripture Trace

He is like a tree planted by streams of water.

Psalm 1:3

Numbered Terraces

Number the terraces in order. Begin at the bottom and write the correct number in each box. Draw some crops growing in the flat areas of the terraces, then color the picture.

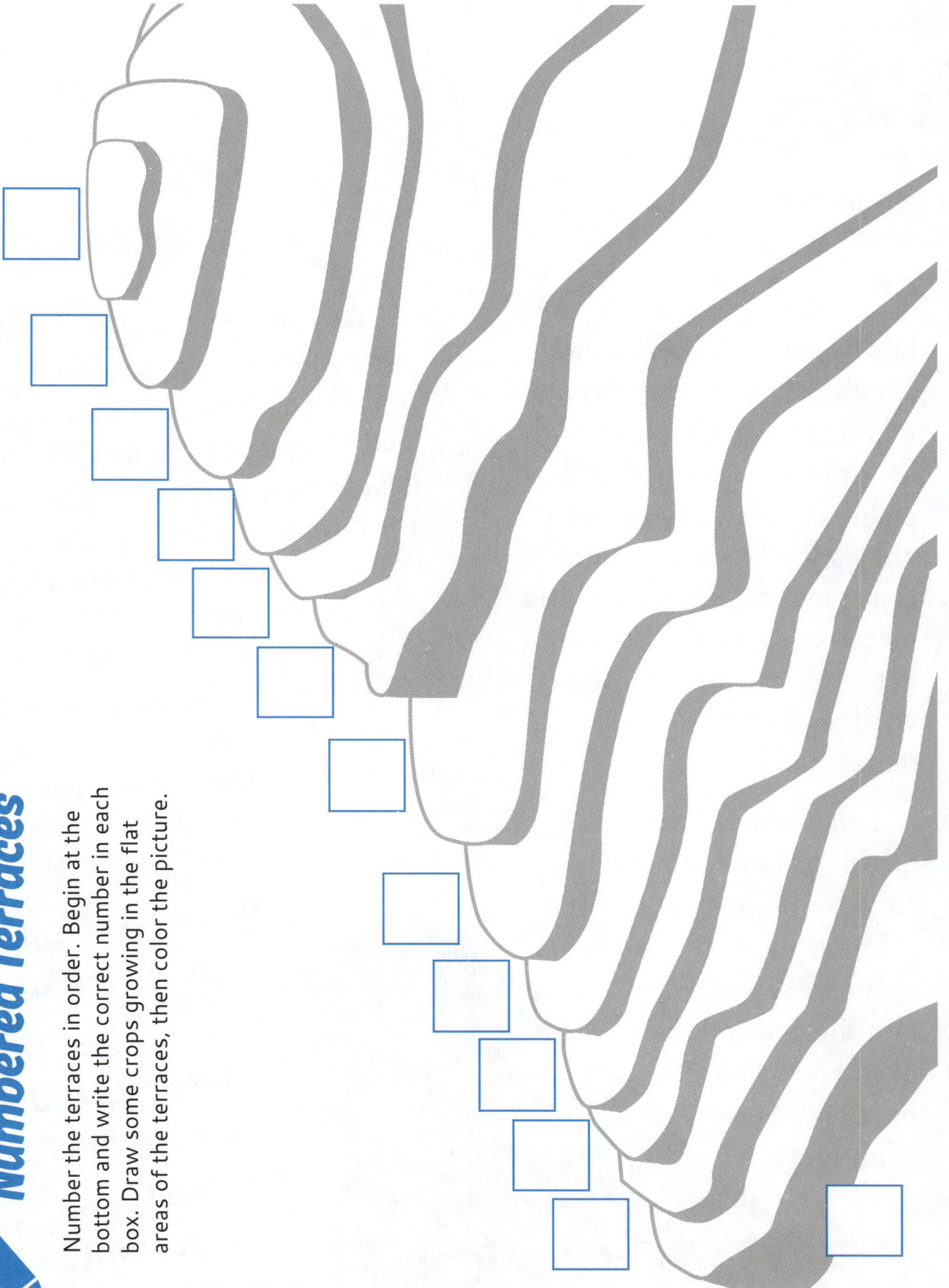

Soil

Plants need soil to grow in.

Soil is made up of bits of sand, clay, and dead plants. The sand and clay are bits that have been broken off of rocks by erosion. When plants die, they dry up and rot. They are broken down into little pieces. All of these little pieces of rocks and plants get mixed together by water, wind, and animals. This makes soil.

Soil is needed for growing plants. It provides food for them and gives the plants a place for their roots to grow. Water soaks into the soil and the plant roots soak it up. Soil is very important.

Looking at Soil

Get two samples of soil — one from your yard and one sample of potting soil. Spread them side by side on a piece of white paper. Look at both with a magnifying glass. The different colors and shapes in the first sample are different bits of rocks and plants. The potting soil is probably darker and might have white pieces in it. Potting soil is made for growing house plants. Both kinds of soil are important for growing different kinds of plants.

? What is soil made from?

? Where do the bits of sand and clay come from?

? Why is soil important?

Things that Make Soil

Find and circle the things that make soil.

ROCKS   ROOTS   WATER   ANIMALS

PLANTS   WORMS   WIND   AIR

```
R W O R M S A O O
O R A T R S L W M
C R M W S D W S N
K A W I W M A R I
S R A N P S T O O
L O D D T A E O A
O E O S N I R T I
A N I M A L S S R
P L A N T S L M T
```

Grand Canyon

One of the most beautiful places on Earth is the Grand Canyon in

Arizona. It is a very deep _canyon_ with steep walls. You can see many layers of sedimentary rock. The Grand Canyon was formed as moving water wore away a huge amount of rock. The canyon was made by erosion.

Some scientists believe that the canyon was made by the river that flows through the bottom. They believe that it took millions of years for the river to wear away the rocks. They believe that a little water and a lot of time made this canyon.

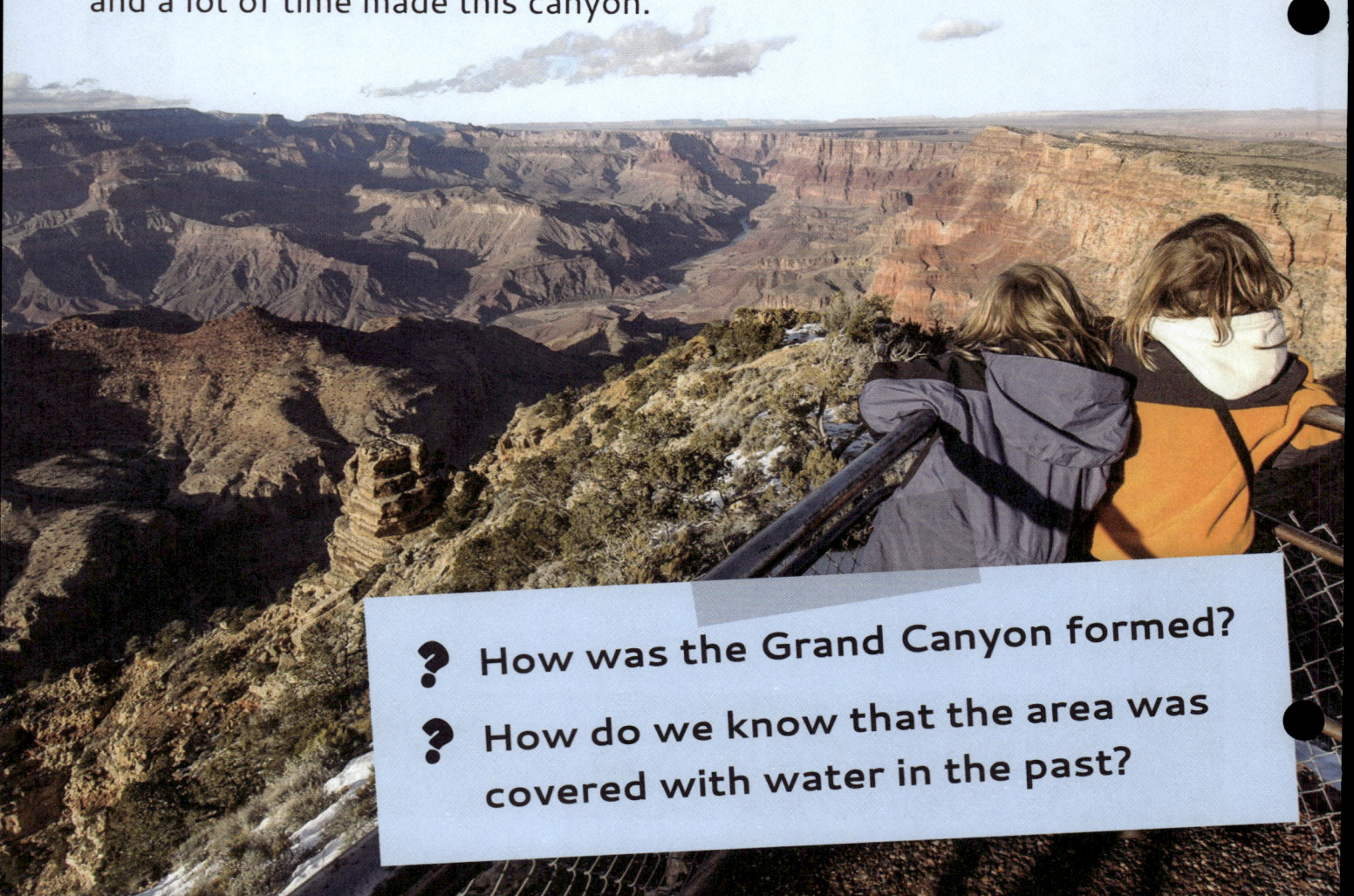

? How was the Grand Canyon formed?

? How do we know that the area was covered with water in the past?

But the Bible gives us the true history of the earth. We know that the whole earth was covered by a giant flood. Creation scientists believe that the Grand Canyon was made by a lot of water in a short period of time. This water came from the Flood.

Many fossils of sea creatures have been found in the different layers of sedimentary rock found in the canyon. These creatures were buried in the Flood. The Bible helps us understand how these rocks and fossils were formed.

Fossils of sea creatures found in the Grand Canyon

Travel Destination

Have someone help you research the Grand Canyon as a travel destination. Use the library or internet. Print out pictures or draw your own pictures. Write down or tell your teacher what you would like to do at the canyon, or why you would like to visit the Grand Canyon.

What Made the Grand Canyon?

Color this picture showing the beauty of the Grand Canyon. Use different colors for the different layers of rock. Then collect and unscramble the letters hidden in the layers to answer the above question.

Water left over from the _____

Caves

A __cave__ is a hollow area underground or in the side of a mountain. Caves are very interesting. Water often flows through the rocks in a cave. If these rocks are made of limestone, the water picks up chemicals from the limestone as it goes through the rocks. As this water dries up, it leaves these chemicals behind. These chemicals build up. They make beautiful formations inside the cave as water slowly drips.

Sometimes the chemical builds up on the ceiling of the cave. This makes rocks that look like icicles. These formations are called __stalactites__.

Other times the chemical builds up on the floor of the cave. These formations are called _stalagmites_. Sometimes a stalactite and a stalagmite grow together to form a _column_.

SEE OPTIONAL ACTIVITY

? What is a cave?

? Where might you find stalactites?

? Where might you find stalagmites?

Rhymes with...

The ends of these two words rhyme — stalac**tite** and stalag**mite**. They both end with _ite_. Tight and might also rhyme. They both end with _ight_.

Below are some clues for other rhyming words with these same endings. Can you fill in the beginning letter or letters of each word?

_ _ ite

_ _ ite Color

_ _ ite

_ _ ight

_ _ ight not wrong

_ _ ight

name _____

Might and Tight

Color this picture. It will help you remember the difference between a stalactite and a stalagmite.

Stalac**tites** hold **tight** to the ceiling.

You **might** trip over a stalag**mite**.

Rock Collection

You can be a rock hound. A rock hound is someone who collects rocks. Back in Lesson 9 you were asked to start collecting rocks. Hopefully you have quite a few by now. In Lesson 16 you learned how to identify your rocks by looking at their colors, shininess, and crystals.

Now it is time to display your rock collection. Use a container with different sections. Sort your rocks any way you like. You could sort them by color or by size. Label where you found each rock. You can use a rocks and minerals guidebook to help you decide what type of rock each one is. Label the rocks you are able to identify. Then show your collection to your friends and family. Other people will enjoy seeing your rock collection.

✏️ Scripture Trace

The earth is the Lord's,
and all its fullness.

Psalm 24:1

Collecting Rocks

Ronny collected some rocks. He sorted them by color and made a chart to show how many of each color he found. Look at his chart to answer the questions.

Color of Rock	Number
Blue	2
Black	5
Red	2
Green	1
Gray	8
Brown	7

1. How many black rocks did Ronny find? _____

2. How many brown rocks did he find? _____

3. Which color of rock did he collect the most of? _____

4. Which color of rock does Ronny have the fewest of? _____

5. He found an equal number of which two colors? _____ and _____

6. How many red and green rocks added together did he find? _____

7. Challenge: How many rocks did Ronny find altogether? _____

Use the blank chart below to sort and count your rocks by color.

Color of Rock	Number

Appreciating Planet Earth

You have learned a lot about the planet Earth. It is the only planet that has life. It has mountains and valleys. It has land and water. It has caves, volcanoes, and glaciers. It is a beautiful place. It was specially designed by God.

✏️ Scripture Trace

And God saw everything
that he had made, and
behold, it was very good.

Genesis 1:31

God Created the Earth

Use the picture clues to complete the crossword about our planet.

Cave

Iceberg

Mountain

Rock

Soil

Water

Unit Vocabulary Review

Draw a line from the vocabulary word with the picture that shows its meaning.

Geyser

Landslide

Avalanche

Terrace

Soil

Canyon

Stalactite

Stalagmite

Column

Optional Activities

Lesson Activities
Requiring an Adult

Weather and Water

NOTE: There are no additional activities requiring an adult for the Water and Weather section of lessons.

Universe

Lesson 11 — Our Solar System — Day 131

SOLAR SYSTEM MOBILE

If you want to have more hands–on work, you can have your student make a mobile of the solar system. Begin with Lesson 12 and add pieces as you go through this unit. Or you could do the whole project at the end of the unit. Cut a sun and planets out of white tagboard/poster board. Trace around a large cereal bowl for the sun. Use other smaller round objects to make the various planets and the moon. Have the students color both sides of each piece using the pictures in each lesson as a guide. Then punch a hole in the top of each circle and use various lengths of string to hang them from a clothes hanger.

Lesson 34 — Final Project — Day 172

FINAL PROJECT — SOLAR SYSTEM MODEL

As an alternative to making a clay model of the solar system, you can purchase inexpensive Styrofoam™ or plastic models that your child can paint.

Planet Earth

Lesson 10 — Igneous Rocks — Day 17

MAKING IGNEOUS ROCKS

You can make your own igneous rocks. First, you need some old crayons. Broken pieces are just fine. Place foil muffin cups in a muffin pan, then place unwrapped crayon pieces in the muffin cups. Place the pan in a 250°F oven until the crayons completely melt. This is just like when rocks melt inside the earth. Then remove the pan from the oven and let the wax cool. This is how igneous rocks are formed. When your "rocks" are cool, remove the foil and you have new igneous rock crayons to use.

Lesson 11 — Sedimentary Rocks — Day 18

EATING SEDIMENTARY ROCKS

You can make sedimentary rocks you can eat. Combine 2 cups of Rice Krispies, ½ cup of candy pieces, and ½ cup of raisins. These are your bits of rock, sand, and sea shells. Now make your glue. Melt 20 large marshmallows and 2 tablespoons of butter in a large pan. Add your rock pieces. Stir it all up. Spread the mixture in a buttered pan. Let your rocks cool. Cut them into squares. Notice how the pieces are all glued together. Now you can eat your sedimentary rocks. (You may also consider dividing the batch in half and stirring 1 tablespoon of cocoa into half of the marshmallow/butter mixture and then spread each half in two separate layers.)

Lesson 14 — Metamorphic Rocks — Day 23

MAKING METAMORPHIC ROCKS

In Lesson 10 you used crayons to make igneous rocks. Today we will use crayons to make metamorphic rocks. Unwrap two different colored crayons and place them beside each other on a piece of aluminum foil. Wrap the foil carefully around the crayons. Use oven mitts to hold the foil. Have an adult help you blow warm air from a hair dryer on the low setting over the crayons. Heat the crayons until they become soft, but not melted. Use the oven mitts to squeeze and twist the foil with the crayons inside. Allow the crayons to cool then unwrap them. You now have a new "metamorphic rock" crayon made by heat from the hair dryer and pressure from your hands.

Lesson 24 — Volcanoes — Day 37

WATCH A VOLCANO ERUPT

Place a bottle on a cookie sheet or in a pan. Form a mountain around the bottle using modeling clay or play dough. Pour 1 teaspoon of baking soda into the bottle. Pour ½ cup of vinegar into the bottle. Watch the volcano erupt.

Lesson 28 — Erosion — Day 43

WATCH EROSION

Make several mud balls. These will be your rocks. Get an adult to help you dry your "rocks." Heat your rocks in an oven set at 275°F for one hour or until they are dry. Allow the rocks to cool. Set your rocks in a tray. Place the tray under a faucet (This is best done outside). Turn the water on so that it drips slowly. Watch as the water slowly wears away small bits of your rocks. Your rocks will break apart much more quickly than regular rocks because they are not as hard.

If you do not have access to dirt to make mud balls, you can do the following similar experiment. Place several inches of sand in a dish. Place the dish in a sink with a pan to catch any sand that falls off. Turn on the water to very slowly drip into the pan. After several minutes, look at the sand. How does it look compared to how it looked to begin with? You will likely see a hole or a trench in the sand. This is because the dripping water has worn away and moved the bits of sand.

Lesson 33 — Caves — Day 51

GROWING "STALAGMITES"

Heat a cup of water until it is boiling. Stir in a teaspoon of salt. The salt "disappears" because it is dissolving in the water just as the chemicals in the cave dissolve in the water. Add two more teaspoons of salt. Drop several large drops of the saltwater on a piece of dark construction paper. Place the paper where it will not be disturbed until all of the water has dried up. When the water is gone, you will see salt crystals left behind. These crystals would be the beginnings of stalagmites if they were in a cave.

Weather and Water
Answer Key

Unit 1: Atmosphere and Meteorology

Lesson 1 — God Made Weather

ANSWER KEY

What do you like to do on a sunny day? **Answers will vary.**

What kinds of things can you do on a rainy day? **Answers will vary.**

What are your favorite things to do on a snowy day? **Answers will vary.**

What are the four seasons? **Spring, summer, fall (or autumn), and winter.**

✏️ FAVORITE SEASON

summer

autumn

spring

winter

Lesson 2 — The Atmosphere

ANSWER KEY

What is the atmosphere? **The air around our planet**

What are the two main ingredients in air? **Nitrogen and oxygen**

What are some good things that our atmosphere does for us? **It keeps the temperatures from getting too hot and too cold, makes weather so rain moves from place to place, provides oxygen, protects us from meteors and other things in space.**

Why does the moon have so many more craters than the earth? **It does not have an atmosphere to protect it.**

Lesson 3 — The Weight of Air

ANSWER KEY

Does air have weight? **Yes**

What do we call air pressing down on things? **Air pressure**

Why is air pressure important? **Because it affects the weather**

✏ AIR HELPS THINGS FLY

Number of things flying: **20**

Lesson 4 — The Study of Weather

ANSWER KEY

What is a meteorologist? **Someone who studies and predicts weather**

What are two things a meteorologist looks at to predict the weather? **What the weather is like today and how the air is moving**

✏ PREDICT THE WEATHER

✏ Unit 1 Vocabulary Review

- Air surrounding the planet
- Person who studies weather
- Main ingredients in air
- Telling what the weather should be like tomorrow
- Air pressing on everything

nitrogen and oxygen

meteorologist

predict

atmosphere

air pressure

Unit 2: Ancient Weather and Climate

Lesson 5 — Weather vs. Climate

ANSWER KEY

If I say that it is raining outside, am I talking about weather or climate? **Weather**

If I say that it is usually cold in the winter, am I talking about weather or climate? **Climate**

Name three different climates. **Polar, tropical, desert, and temperate were mentioned in the lesson.**

✏ CLIMATE MATCH WORKSHEET

Lesson 6 — Climate Before the Flood

ANSWER KEY

What clues does the Bible give about the climate before the Flood? **It was warm enough that Adam and Eve did not need to wear clothes.**

What clues do fossils give us about the climate before the Flood? **Fossils of tropical plants have been found in every part of the world. This indicates that the climate over the whole world was warmer than it is today.**

✏ GARDEN OF EDEN WORKSHEET

Tropical

Lesson 7 — The Great Flood

ANSWER KEY

Why did God send the Great Flood? **To punish mankind for being so sinful**

Who was killed by the Flood? **Every person and land animal that was not on the ark**

Who was saved from the Flood? **Noah**

and his family and all the animals on the ark

What promise did God make when he sent the rainbow? *He would never flood the entire world again*

What are the colors of the rainbow? *Red, orange, yellow, green, blue, indigo, and violet*

Lesson 8 — Climate After the Flood

ANSWER KEY

Why was there an ice age after the Flood? *The climate changed. There was a lot more snowfall and it was a lot colder in many areas.*

What happened to all the ice from the Ice Age? *Most of it melted and flowed into the oceans. Some of it is still in the glaciers near the North and South Poles.*

✏️ Unit 2 Vocabulary Review

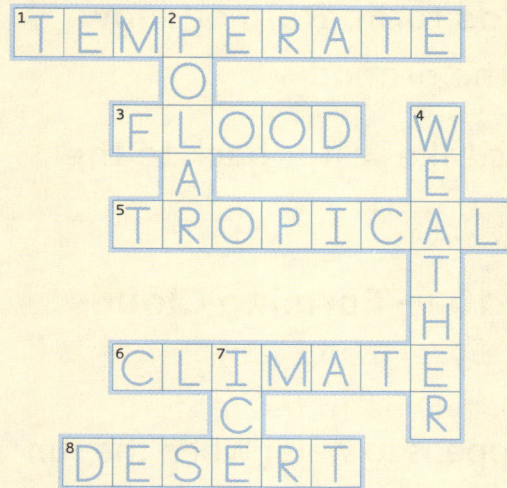

Unit 3: Clouds

Lesson 9 — Water Cycle

ANSWER KEY

How does water get from the ocean into the air? *The sun heats the water, causing it to evaporate.*

How does water get from the sky to the land? *The wind moves the moist air away from the ocean, then when it cools the water falls to the ground* as rain or snow.

How does water get from the land to the ocean? *It flows from the land into rivers and then into the ocean.*

Why is the water cycle important? *It allows us to use the same water over and over so that we don't run out.*

🖊 THE WATER CYCLE

1. When water is warmed by the **sun,** it evaporates. It **rises** into the air.

2. **Wind** blows the air to the land.

3. **Clouds** form. **Rain** or snow falls to the ground.

4. Rivers take water back to the **ocean.**

Lesson 10 — Forming Clouds

ANSWER KEY

What happens to water when the sun shines on it? ***Some of it turns into a gas and goes into the air.***

How can a cloud float? ***Warm air rises and pushes up on the clouds.***

Lesson 11 — Cloud Types

ANSWER KEY

How does a stratus cloud look? ***Like a solid sheet of clouds***

How does a cumulus cloud look? ***Big and fluffy***

How does a cirrus cloud look? ***Wispy and curly***

Lesson 12 — Precipitation

What is precipitation? ***Water falling from a cloud***

Name three types of precipitation. ***Rain, snow, hail***

How is hail formed? ***Water drops are pushed up to the top of a cloud where it is cold. They freeze. As they fall back down more water collects on the ice. It is pushed back up to the top of the cloud. It freezes again. This continues until the ice becomes too heavy to be pushed back up.***

🖊 Unit 3 Vocabulary Review

ANSWER KEY

1. Water is reused through the **water cycle.**

2. Water **evaporates** when it turns into a gas.

3. **Condensation** happens when water turns into a cloud.

4. A **cloud** is formed when water turns into a liquid in the sky.

5. A cloud can **float** because warm air pushes up on it.

6. **Stratus** clouds form a big sheet of clouds in the sky.

7. **Cumulus** clouds are big and fluffy.

8. **Cirrus** clouds are curly and light.

9. Water falling from the sky is called **precipitation**.

10. **Rain** is the most common form of precipitation.

11. **Snow** forms when water turns into ice crystals in a cloud.

12. **Hail** forms when water is pushed up into a cold part of a cloud over and over again.

Unit 4: Storms

Lesson 13 — Air Masses and Weather Fronts

ANSWER KEY

What is an air mass? *A large amount of air that is all the same temperature*

What is a weather front? *Where two air masses touch*

What happens at a weather front? *The air gets jumbled and wind and storms can happen*

Lesson 14 — Wind

ANSWER KEY

What is wind? *Air that is moving quickly*

What causes the wind to blow? *The air over land heats up more quickly than air over the water. The cooler air moves into areas with warmer air. This moving air is the wind.*

Why is wind important? *Wind moves air masses causing weather fronts that bring rain.*

🖉 USING WIND WORKSHEET

Lesson 15 — Thunderstorms

ANSWER KEY

Are thunderstorms more likely to happen in the summer or the winter? *It depends where you live. If you live in a tropical area, they can happen any time. If you live in a temperate area, they are more common in the summer. Thunderstorms need warm, moist air.*

What causes lightning? *Water drops rubbing against each other causes a build up and release of electricity inside a cloud, which causes a flash of light.*

What causes thunder? *Lightning heats the air around it. The hot air moves outward quickly and causes a loud sound.*

Where should you be during a thunderstorm? *Inside a building*

Lesson 16 — Tornadoes

ANSWER KEY

What is a tornado? *A swirling storm cloud that sucks things up like a vacuum cleaner*

When do most tornadoes occur? *In the spring*

Where do most tornadoes occur? *In the central and eastern parts of the United States*

Where is the safest place to be during a tornado? *In a basement, under the stairs, under a strong table, or in a bathtub with cushions on top of you*

Lesson 17 — Hurricanes

ANSWER KEY

What are the largest storms in the world called? *Hurricanes*

Where do hurricanes usually form?

● **Near the equator over warm water**

What is the center of a hurricane called? **The eye**

✏️ **Unit 4 Vocabulary Review**

ANSWER KEY

Air mass

Weather front

Wind

Thunderstorm

Lightning

Thunder

Tornado

Hurricane

Unit 5: Weather Information

Lesson 18 — Measuring Temperature and Air Pressure

ANSWER KEY

What instrument do we use to measure temperature? **Thermometer**

What instrument do we use to measure air pressure? **Barometer**

Why do meteorologists look for changes in air pressure? **So they can know where weather fronts are**

Lesson 19 — Measuring Rainfall and Wind Speed

ANSWER KEY

What instrument is used to measure rainfall? **Rain gauge**

What instrument is used to measure wind speed? **Anemometer**

What does an anemometer look like? **A pole with small cups attached**

✏️ RECORDING RAIN WORKSHEET

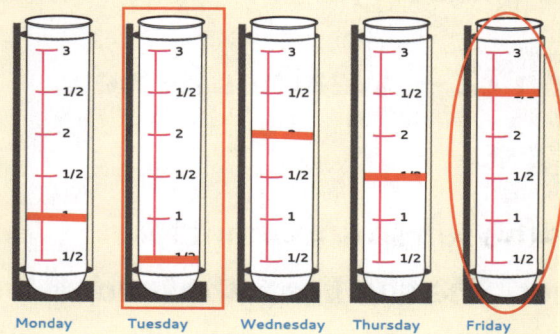

Monday   Tuesday   Wednesday   Thursday   Friday

Lesson 20 — Predicting Weather

ANSWER KEY

How do scientists get weather

information out on the ocean? **They have instruments on ships**

How do scientists get weather information from high in the sky? **They use weather balloons and airplanes**

Where do meteorologists send all of their weather information? **The National Weather Service**

What do meteorologists do with their weather maps? **Use them to make weather predictions**

Lesson 21 — Weather Sayings

ANSWER KEY

What is your favorite weather saying? **Answers will vary.**

Lesson 22 — Weather Review

ANSWER KEY

Why is the sun so important for weather? **The sun heats the land, water, and air. This causes air to move and water to evaporate, which creates weather.**

What temperature of air rises? **Hot**

What temperature of air falls? **Cold**

What do we call moving air? **Wind**

What happens when the air gets so full of heavy water drops that it cannot hold all of it? **The water falls to the ground as rain, snow, or hail.**

✏ Unit 5 Vocabulary Review

thermometer   air pressure

rain gauge   wind speed

barometer   temperature

anemometer   rainfall

National Weather Service

Unit 6: Ocean Movements

Lesson 23 — Oceans

ANSWER KEY

Which ocean is the biggest? **Pacific Ocean**

Which ocean is on the east side of the United States? **Atlantic Ocean**

Which ocean is near the south pole? **Southern Ocean**

Can you name all five oceans without looking? **Atlantic, Pacific, Indian, Arctic, Southern**

✏ OCEANS WORKSHEET

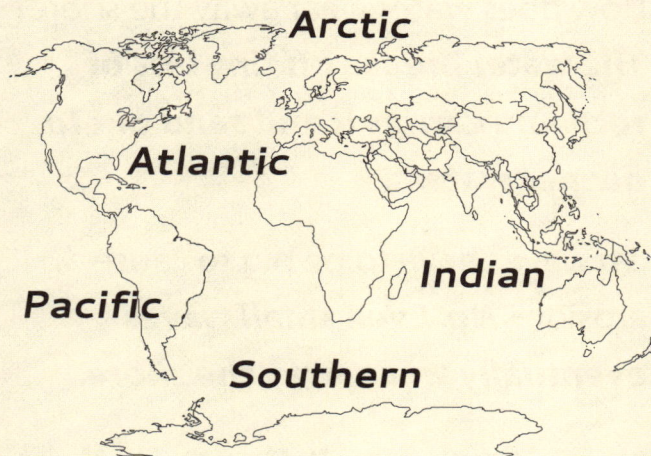

Lesson 24 — Why is Seawater Salty?

ANSWER KEY

How does salt get into the ocean? **Water dissolves salt when it flows across land then carries it to the oceans.**

How does water leave the ocean? **By evaporation**

Why does salt stay in the ocean? **When the water evaporates it leaves the salt behind.**

Lesson 25 — Ocean Currents

ANSWER KEY

Is the water warmer near the equator or near the North Pole? **Warmer near the equator**

What direction does cooler water move? **From the poles toward the equator**

What do we call moving rivers in the ocean? **Currents**

Lesson 26 — Waves

ANSWER KEY

What causes a wave to form? **Wind pushes water, causing it to move and push on other water.**

Why do waves usually get taller near the shore? *The bottom of the wave slows down when it hits the ocean bottom near the shore. This causes the top of the wave to pile up and get taller.*

What do we call tall waves falling onto the shore? *Breakers*

Lesson 27 — Tides

ANSWER KEY

What causes the ocean level to change? *Gravity from the moon pulls on the water.*

How many high tides are there each day? *Two*

Why did God design the ocean water to move? *Moving water keeps the oceans clean and keeps temperatures move even.*

✎ **TIDES WORKSHEET**

Low tide

High tide

Lesson 28 — Wave Erosion

ANSWER KEY

How does water wear away the shore? *The water breaks off tiny bits of rock and carries it and sand out to deeper waters.*

Do waves have to be big to cause erosion? *No. Even small waves eventually wear away the shore.*

Lesson 29 — Building Beaches

ANSWER KEY

How do waves create new sand? *Waves bring sand from other parts of the ocean and drop it at the*

shore. Waves break off tiny bits of rock and shells that are added to the sand on the shore.

How do rivers create new sand? ***Rivers carry tiny bits of rock and sand from the land and drop it at the shore.***

Unit 6 Vocabulary Review

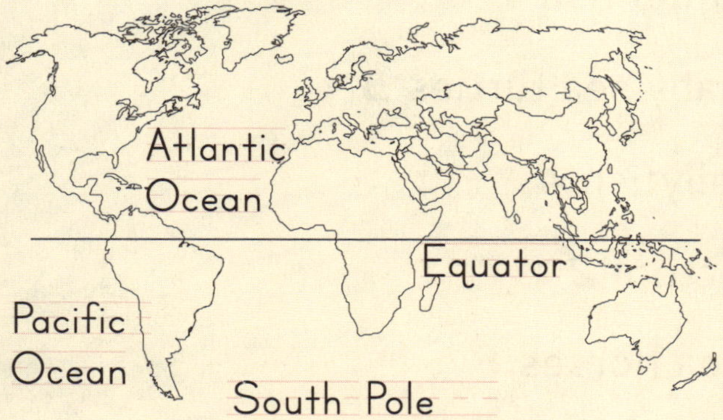
North Pole
Atlantic Ocean
Equator
Pacific Ocean
South Pole

Q	W	A	V	E	S	O	M	R	R
G	P	H	I	G	H	T	I	D	E
P	J	E	Q	W	U	G	G	Q	S
B	E	O	E	R	O	S	I	O	N
C	L	O	W	T	I	D	E	T	H
C	U	R	R	E	N	T	S	Y	D
O	O	L	E	W	R	B	N	U	Z
V	L	B	R	E	A	K	E	R	S
L	W	I	C	H	Z	L	Y	R	L
I	D	D	I	S	S	O	L	V	E

Unit 7: Seafloor

Lesson 30 — Sea Exploration

ANSWER KEY

What can a diver wear to explore the ocean? ***A diving suit and air tank***

What might a diver see in the ocean? ***Many different kinds of fish, coral, seaweed, sharks, dolphins, and other animals***

Why can't divers go very deep into the ocean? ***The deeper you go the harder the water pushes down on you.***

How do people explore deeper in the ocean? ***In a submarine called a submersible***

What have people seen near the bottom of the ocean? ***Sperm whales, glow-in-the-dark creatures, and living sponges***

✏ **SEA EXPLORATION COLORING SHEET**

Living things: *diver, fish, algae, coral, whale, octopus*

Lesson 31 — The Ocean Floor

ANSWER KEY

What is the name of the part of the ocean floor that gently slopes away from the shore? **Continental shelf**

What is the continental slope? **The steep drop-off at the edge of the continental shelf**

What is the name of the flat part of the ocean floor? **Abyssal plain**

Lesson 32 — Ocean Zones

ANSWER KEY

Why do most sea creatures live in the sunlit zone? **This is where plants grow. Most animals either eat the plants or eat the animals that eat the plants, so they have to live close to the plants.**

Name one animal you might find in the twilight zone. **Sperm whale, octopus, sponges, and lantern fish**

were mentioned in the lesson.

Why are there no plants in the midnight zone? **Plants must have sunlight to grow and there is no sunlight in the midnight zone.**

✏ **COUNTING SEA CREATURES WORKSHEET**

angel fish **7**

baby sea turtles **5**

jellyfish **4**

octopi **2**

sea horses **8**

sharks **3**

shrimp **10**

sponges **6**

starfish **9**

whale **1**

Lesson 33 — Vents and Smokers

ANSWER KEY

What is an ocean vent? **An opening in the ocean floor where hot water pours out**

● What mineral is found in this hot water? **Sulfur**

How can animals live near vents when no plants or algae grow there? **Bacteria can eat the sulfur from the vents, then the animals eat the bacteria.**

✎ OCEAN VENTS WORKSHEET

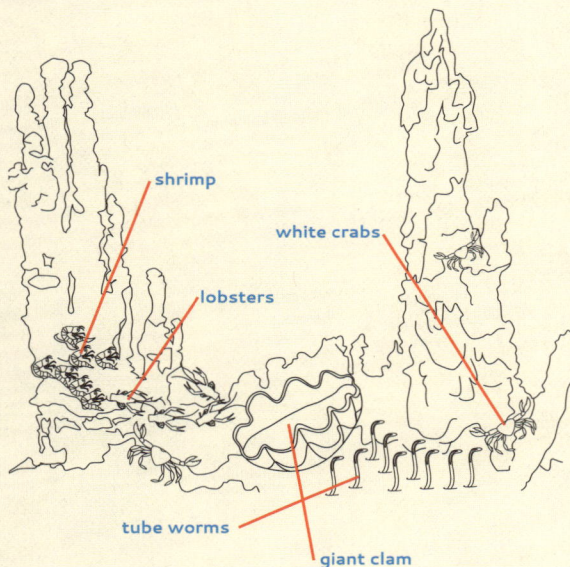

shrimp

white crabs

lobsters

tube worms

giant clam

Lesson 34 — Coral Reefs

ANSWER KEY

What is a coral reef? **An underwater island formed by coral**

How is a coral reef formed? **Each coral builds a hard cup around itself. When it dies, a new coral builds on top of the empty cup. Over a long period of time the reef grows bigger and bigger.**

Where are most coral reefs found? **In warm, clear water near the equator**

Lesson 35 — Conclusion

ANSWER KEY

What is the most interesting thing you learned about weather? **Answers will vary.**

What is the most interesting thing you learned about the ocean? **Answers will vary.**

✎ **Unit 7 Vocabulary Review**

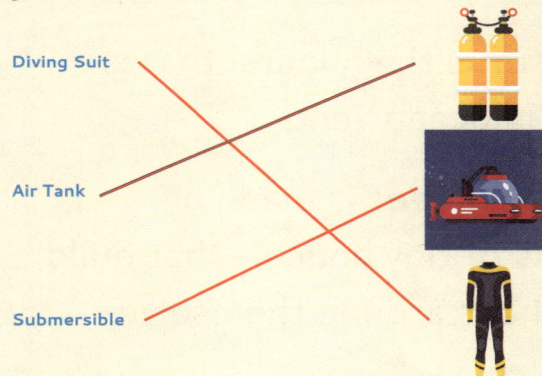

Diving Suit

Air Tank

Submersible

Continental Shelf

Continental Slope

Abyssal Plain

Midnight Zone

Sunlit Zone

Vent

Twilight Zone

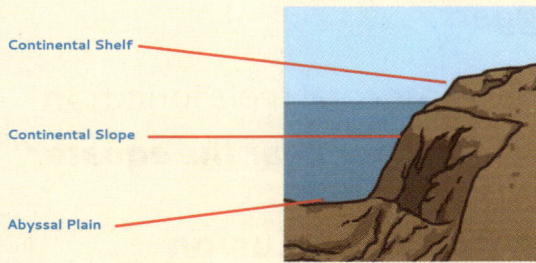

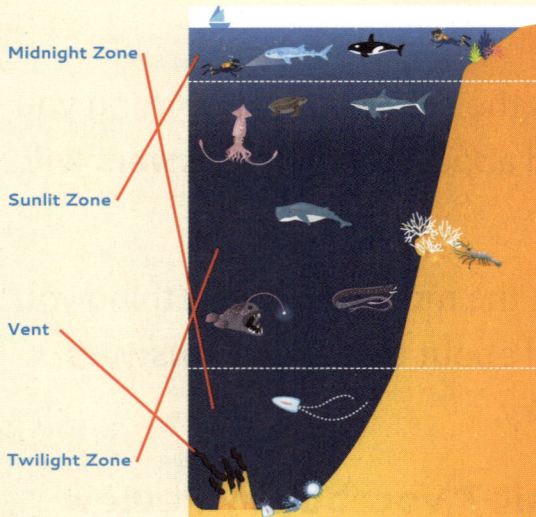

1. **Sulfur** is the mineral that is found in ocean vents.

2. A **coral reef** is made by millions of tiny animals that build hard shells around themselves.

Universe
Answer Key

Unit 1: Space Models and Tools

Lesson 1 — Introduction to Astronomy

ANSWER KEY

What is astronomy? **The study of space**

What questions do you have about space? **Answers will vary.**

✏️ **IN THE BEGINNING WORKSHEET**

sun — Created on day **4**

the earth (dry land and seas) — Created on day **3**

moon — Created on day **4**

water animals — Created on day **5**

stars — Created on day **4**

land animals — Created on day **6**

light — Created on day **1**

the first man — Created on day **6**

Lesson 2 — The Earth Is Moving

ANSWER KEY

How long does it take for the earth to go around the sun one time? **One year**

How long does it take for the earth to spin around one time? **One day/24 hours**

What do we call the spinning motion of the earth? **Rotation**

What force keeps all of the planets moving around the sun? **Gravity**

✏️ **DAY AND NIGHT WORKSHEET**

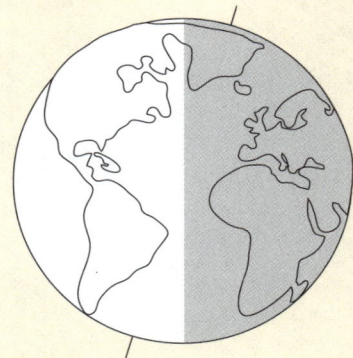

Day   Night

Lesson 3 — Why Do We Have Seasons?

ANSWER KEY

What are the four seasons? **Spring, summer, fall, winter**

Why do we have different seasons? **The earth is tilted so the sun shines differently on the earth** *throughout the different parts of the year. This causes the temperature and weather to change.*

Lesson 4 — Telescopes

ANSWER KEY

Why do people like to use a telescope? *It makes things in space look bigger. They can see things better with a telescope.*

How is a telescope like a magnifying glass? *They both make things look bigger. They both use lenses.*

✏️ WHAT CAN I SEE WITH A TELESCOPE?

✏️ **Unit 1 Vocabulary Review**

Astronomy	The study of space ~~The study of water~~
Orbiting	~~The earth spinning~~ The earth moving around the sun
Rotation	The earth spinning ~~The earth moving around the sun~~
Gravity	Force pulling down on things ~~Force pushing things away~~
Seasons	~~Salt, pepper, cinnamon, nutmeg~~ Summer, winter, spring, fall
Telescope	Instrument used to see things that are far away ~~Instrument used to see things that are very small~~

Unit 2: Outer Space

Lesson 5 — Overview of the Universe

ANSWER KEY

How big is the universe? *No one really knows. It is gigantic.*

Why is the sun a special star? *It is the closest star to Earth. It gives light and heat to Earth.*

What is the solar system? *The sun and the planets and moons that move around it.*

Lesson 6 — Stars

ANSWER KEY

What are stars made of? **Super-hot gas**

Why is our sun just right for life on Earth? **It is the right distance, size, and brightness to give us the light and heat that we need. It is not too hot or too cold, too bright or too dim.**

Are all stars the same size and brightness? **No**

✏️ STAR WORKSHEET

Lesson 7 — Our Galaxy

ANSWER KEY

What is a galaxy? **A large group of stars that all spin around a central point**

What is the name of the galaxy we live in? **The Milky Way**

What are two different shapes of galaxies? **Oval and spiral were mentioned in the lesson.**

✏️ MILKY WAY MAZE

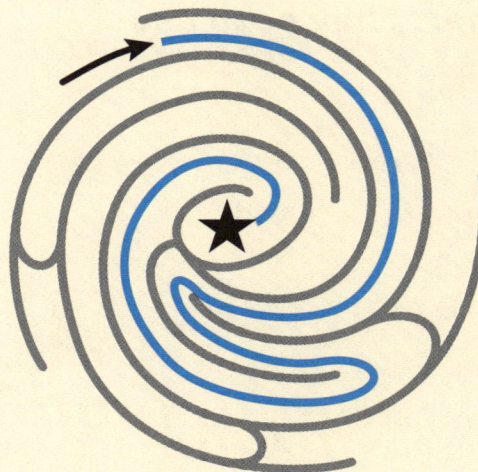

Lesson 8 — Asteroids

ANSWER KEY

What is an asteroid? **A large rock that orbits the sun**

Most asteroids are found between which two planets? **Mars and Jupiter**

What is this area called? **The asteroid belt**

What is the name of the largest asteroid? **Ceres**

Lesson 9 — Comets

ANSWER KEY

What is a comet made of? **Mostly ice with some dust and bits of rock frozen in it**

What happens when a comet gets close to the sun? **It begins to melt and turns to gas. The gas and dust are pushed back away from the sun giving the comet a tail.**

What are the two parts of a comet? **Head and tail**

Lesson 10 — Meteors

ANSWER KEY

What is a meteor? **A rock that has gotten too close to Earth and is pulled down by Earth's gravity.**

Why does a meteor become a shooting star? **It burns up as it goes through the atmosphere. This gives off a bright light.**

✏ Unit 2 Vocabulary Review

1. The **universe** contains billions of stars.

2. Our **solar** system is made up of all of the planets and moons that orbit the sun.

3. A **constellation** is a group of stars that forms a picture.

4. The **sun** is the closest star to the earth.

5. A **galaxy** is a large group of stars that all move together.

6. Our galaxy is called the **Milky Way**.

7. An **asteroid** is a large rock that orbits the sun.

8. Most asteroids are located in the asteroid **belt**.

9. A **comet** is a ball of ice, dust, and rock that orbits the sun.

10. The bright part of a comet is the **head**.

11. The dust and gas that spread out from a comet form its **tail**.

12. A **meteor** is a rock that burns up in the earth's atmosphere.

Unit 3: Sun and Moon

Lesson 11 — Our Solar System

ANSWER KEY

Name three things that are in our solar system. *Sun, planets, moons, asteroids, comets, dwarf planets were mentioned in the lesson.*

How are the four planets that are closest to the sun different from the four planets that are farthest away? *The closer planets are smaller and the farther ones are bigger. In later lessons the student will learn that the closer planets are solid and the further planets are gas.*

Name two dwarf planets. *Ceres, Pluto, Eris were mentioned in the lesson.*

Lesson 12 — Our Sun

ANSWER KEY

Which is larger, the sun or the earth? *The sun is much larger than the earth.*

How big is our sun compared to other stars? *It is a medium-sized star.*

Why do we need the sun? *It gives light and heat to the earth. It allows plants and animals to live.*

✏️ COLORS OF LIGHT

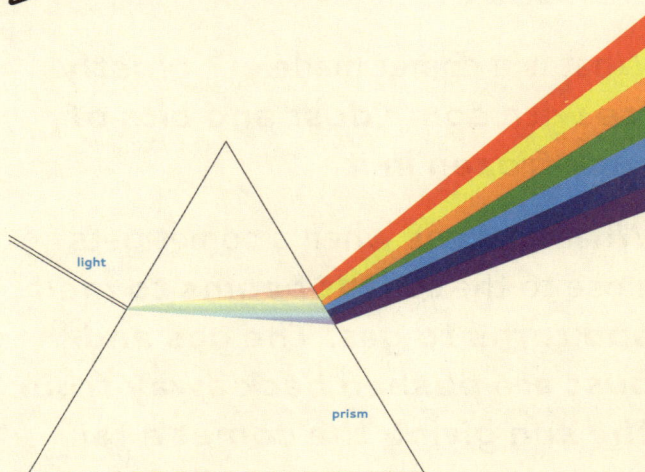

light

prism

Lesson 13 — The Surface of the Sun

ANSWER KEY

Why should you never look directly at the sun? *It is so bright it will damage your eyes.*

What is a sunspot? *A cooler area on the surface of the sun*

What is a solar flare? *An explosion on the surface of the sun*

Lesson 14 — Solar Eclipse

ANSWER KEY

What causes a solar eclipse? *The moon moves directly between the earth and the sun and blocks the light from the sun.*

During an eclipse is the whole earth dark? *No. An area only about 150 miles wide is dark.*

Lesson 15 — Solar Energy

ANSWER KEY

What is solar energy? *Energy that comes from the sun*

What types of energy come from the sun? *Heat and light*

What are two things that people have invented to help them use solar energy? *Solar panels and solar cells*

✏ SOLAR ENERGY WORKSHEET

Lesson 16 — Our Moon

ANSWER KEY

How long does it take for the moon to go around the earth one time? *About one month*

Where does the light from the moon come from? *It is reflected from the sun.*

Why are there no plants or animals living on the moon? *There is no air/atmosphere or water on the moon.*

Lesson 17 — Phases of the Moon

ANSWER KEY

Why does the light shining on the moon change from day to day? *The* moon is in a different position between the earth and the sun each day because it is moving around the earth.

What is a full moon? *When light shines on Earth from the whole surface of the moon*

What is a new moon? *When no light shines on Earth from the moon*

Lesson 18 — Where Did the Moon Come From?

ANSWER KEY

Where did the moon come from? *God created it by speaking it into existence*

On which day of creation did God make the moon? *Day 4*

WHAT IS TRUE ABOUT THE MOON?

Unit 3 Vocabulary Review

Sunspots	Cool areas on the sun's surface
Solar Flare	Moon covers the sun
Solar Eclipse	Explosion on the sun's surface
Solar Panel	Creates electricity from sunlight
Solar Cells	Heats water using sunlight
Moon	How sunlight bounces off the moon
Reflect	When no sunlight reflects off of the moon
Full Moon	Large rock that orbits the earth
New Moon	When the whole surface of the moon is lit up

Unit 4: Planets

Lesson 19 — Mercury

ANSWER KEY

Why would it be a bad idea to visit Mercury? **It has no atmosphere so it is either very hot or very cold. There is no safe temperature on Mercury.**

Is Mercury closer to the sun or farther away from the sun than Earth is? **It is closer. It is the closest planet to the sun.**

✏️ MERCURY FACTS SHEET

Mercury is the **closest** planet to the sun.

Mercury is the **smallest** planet.

Mercury has no **atmosphere**.

Lesson 20 — Venus

ANSWER KEY

Why is Venus so hot? **It has thick poisonous clouds that trap the sun's heat.**

Why can't people live on Venus? **There is no oxygen, so we would not be able to breathe. It is too hot; we would burn up.**

Why is Venus called the Morning Star? **Sunlight reflects off of the clouds. This makes it bright in the night sky. Venus is often the first thing to appear in the night sky and the last thing you can see before the sun comes up.**

✏️ VENUS FACTS SHEET

Venus is just a little smaller than **Earth**.

Venus is the **second** planet from the sun.

Venus is the **hottest** planet in the solar system.

The atmosphere on Venus is **poisonous** to humans.

Lesson 21 — Earth

ANSWER KEY

List three ways that Earth is just right for people. **Possible answers include: It is the right distance from the sun, it has an oxygen atmosphere, it has lots of water, it has plants to produce the oxygen and provide food, the clouds move water around the** *earth, the moon gives light at night.*

Which planets are closer to the sun than Earth? *Mercury and Venus*

✏️ EARTH FACTS SHEET

Earth is the **third** planet from the sun.

Earth is designed just right for **people** to live there.

Earth has a lot of **water**.

Earth has air with **oxygen**.

Lesson 22 — Mars

ANSWER KEY

What is the name of the fourth planet from the sun? **Mars**

Why is Mars sometimes called the Red Planet? **The soil has rust in it. This makes it look red from Earth.**

Why would you not be able to breathe on Mars? **The air is mostly carbon dioxide. There is not enough oxygen.**

✏️ MARS FACTS SHEET

Mars is the **fourth** planet from the sun.

Mars is a little bigger than **Mercury**.

Mars is called the **Red** Planet.

Lesson 23 — Jupiter

ANSWER KEY

How is Jupiter different from Earth? **Possible answers include: It is much larger. It is made of gas instead of rock. It is farther away from the sun.**

What is the Great Red Spot? **A giant storm on Jupiter**

✏️ JUPITER FACTS SHEET

Jupiter is the **largest** planet in the solar system.

Jupiter is the **fifth** planet from the sun.

Jupiter is made of **gas**.

Jupiter has over 70 **moons**.

The Great Red Spot is a giant **storm**.

Lesson 24 — Saturn

ANSWER KEY

How many moons does Saturn have? **At least 80**

What are Saturn's rings made of? **Bits of ice, dust, and rock**

✏️ SATURN FACTS SHEET

Saturn is the **sixth** planet from the sun.

Saturn is a **gas** planet.

Saturn has thousands of **rings**.

Titan is the largest **moon** that orbits Saturn.

Lesson 25 — Uranus

ANSWER KEY

How does Uranus move in a way that is different from the other planets? **It rolls around sideways compared to how the other planets spin.**

Is Uranus a gas planet or a rock planet? **Gas**

✏️ URANUS FACTS SHEET

Uranus is the **seventh** planet from the sun.

Uranus is a pale **blue** color.

Uranus is a **gas** planet.

Uranus has a few **rings** and at least 20 moons.

Lesson 26 — Neptune

ANSWER KEY

Why does the sun look like a bright star from Neptune? **It is very far away from the sun.**

How big is Triton? **About the same size as our moon.**

✏️ NEPTUNE FACTS SHEET

Neptune is the **farthest** planet from the sun.

Neptune is about the same size as **Uranus**.

Neptune looks **blue**, like Uranus.

Neptune is a **gas** planet.

Triton is Neptune's largest moon.

Lesson 27 — Pluto and Eris

ANSWER KEY

Why are Pluto and Eris called dwarf planets rather than just planets? **They are much smaller than the regular planets.**

Why are Pluto and Eris very cold? **They are very far from the sun, so the sun does not give them any heat.**

Which dwarf planet is farthest from the sun? **Eris**

✏️ Unit 4 Vocabulary Review

Write the names of two dwarf planets that orbit very far from the sun.

Pluto and ***Eris***

Unit 5: Space Program

Lesson 28 — NASA

ANSWER KEY

What is NASA? *A group of people who work on things to help us understand space.*

Name three types of things that NASA does. *Possible answers include: They build spaceships and rockets. They build space probes. They operate the International Space Station and the Hubble Telescope. They train astronauts.*

What is the name of the special telescope that NASA built in space?
The Hubble Space Telescope

✏️ WHAT DOES NASA DO?

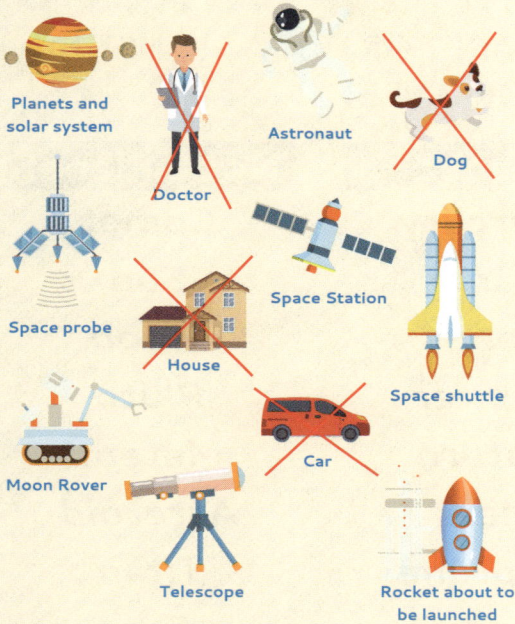

Planets and solar system

Doctor

Astronaut

Dog

Space probe

Space Station

House

Space shuttle

Moon Rover

Car

Telescope

Rocket about to be launched

Lesson 29 — Space Exploration

ANSWER KEY

What is a rocket? *A very powerful engine that shoots hot gas out the end in order to push something very fast.*

What is a satellite? *Something that orbits the earth*

What are some things that have been sent into space by rockets? *Possible answers include: satellites, space probes, astronauts/ people, space shuttle, space station parts, Hubble Telescope.*

READING SIMPLE BIOGRAPHIES OF SOME OF THE EARLY ASTRONAUTS WOULD BE A

GREAT ADDITION TO THIS UNIT.

Lesson 30 — Apollo Program

ANSWER KEY

What was the Saturn V? *The rocket designed to send men to the moon*

What was the job of the Lunar Module? *To take astronauts from the space capsule to the surface of the moon and back*

What vehicle was used to explore the moon? *Lunar Rover*

Lesson 31 — The Space Shuttle

ANSWER KEY

What was the space shuttle designed to do? *Carry people and instruments into space near Earth*

What are the three parts of the space shuttle? *Orbiter, liquid fuel tank, solid fuel rocket boosters*

Lesson 32 — International Space Station

ANSWER KEY

What is the purpose of the International Space Station? *To provide a place for astronauts* to conduct experiments in space for several months at a time

How does the space station get power? *Solar panels convert sunlight into electricity.*

Lesson 33 — Astronauts

ANSWER KEY

What are two things you need to study in school if you want to be an astronaut? **Math and science**

What are three things that a space suit provides that are missing in space? **Air, water, heating and cooling**

Lesson 34 — Final Project — Solar System Model

ANSWER KEY

Which is your favorite planet? **Answers will vary.**

What did you like learning the most in this book? **Answers will vary.**

Lesson 35 — Conclusion

✏️ PLANET OR NOT A PLANET WORKSHEET

Planet	Not a Planet
Mercury	Comet
Earth	Sun
Venus	Moon
Jupiter	Pluto
Saturn	Meteor
Mars	Asteroid
Neptune	Titan
Uranus	Star

✏️ **Unit 5 Vocabulary Review**

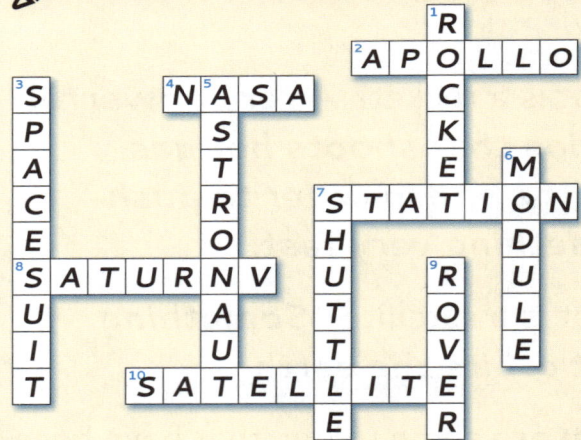

Planet Earth
Answer Key

GOD'S DESIGN®

Unit 1: Origins and Glaciers

Lesson 1 — Introduction to Earth Science

ANSWER KEY

What is earth science? *Learning about the planet Earth*

Where did the earth come from? *God created it.*

What other things did God create? *All of the planets, stars, sun, moon, everything on the earth*

Lesson 2 — The Earth Is Special

ANSWER KEY

What are some things in your house that are made of metal? *Answers will vary, but could include spoons, refrigerator, washer and dryer, door hinges, etc.*

What are some things in your house that are made of plastic? *Answers will vary, but could include cups, hangers, toys, etc.*

What are some things in your house that are made of wood? *Answers will vary, but could include furniture, floors, cutting board, picture frames, etc.*

Why is Earth a special planet? *God created it. It has lots of water and is the right distance from the sun. It has many useful things in it such as soil, metals, and oil.*

Lesson 3 — The Earth's History

ANSWER KEY

What three events did you learn about that have changed the earth? *Creation, Fall, Flood*

Does the Bible tell us the earth is young or old (thousands or millions of years)? *Young — about 6,000 years old*

✏️ EVENTS WORKSHEET

Lesson 4 — The Genesis Flood

ANSWER KEY

Why did God send the Great Flood? *To punish all of the evil people*

What is the ark? *A giant boat that Noah built to save his family and many animals*

How did the Flood change the surface of the earth? *It covered the whole earth with mud and water, forming fossils and new rocks. It also cut away large areas of rock.*

Were there dinosaurs on the ark? *Yes. God said all land animals would be saved on the ark, so a pair of each kind of dinosaur must have been on the ark.*

✏ AFTER THE GREAT FLOOD

There are 11 pairs of animals (5 in the foreground and 6 in the background including the birds in the sky). There are a total of 23 animals (22 in pairs plus the pterodactyl above the ark).

Lesson 5 — The Great Ice Age

ANSWER KEY

How were things different on the earth after the Flood? *The land was cooler. There were more clouds. Volcanoes put ash in the air.*

Why did the ice last so long during the Ice Age? *Some areas were cold enough that ice did not melt completely each year so new ice and snow continued to build up.*

✏ ICE AGE ANIMALS

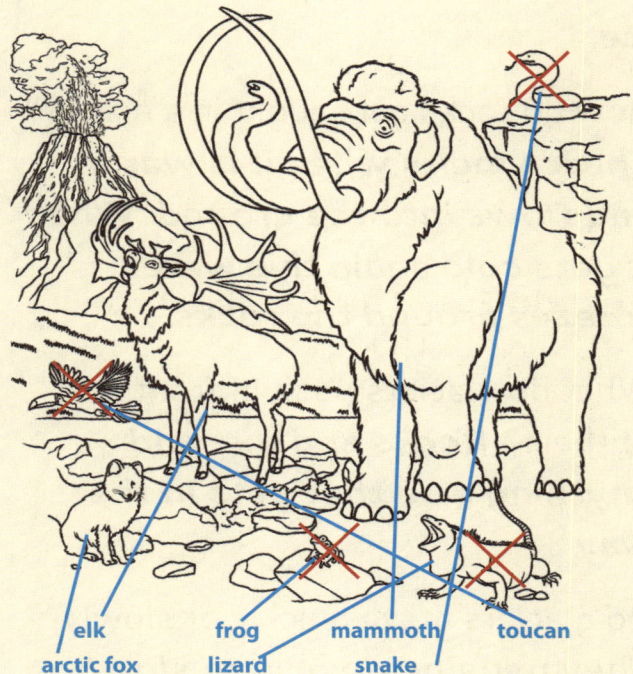

elk   frog   mammoth   toucan
arctic fox   lizard   snake

Lesson 6 — Glaciers

ANSWER KEY

What is a glacier? *A moving sheet of ice that does not completely melt in the summer*

Where are most glaciers found today? *Near the North and South Poles*

What is an iceberg? *A chunk of ice that has broken off of a glacier and is floating in the water*

Lesson 7 — Movement of Glaciers

ANSWER KEY

What makes glaciers move down a mountain? *Gravity pulls on the ice.*

How do rocks get caught in a glacier? *The ice melts when it is warm and flows into the ground. When it gets cold again this water freezes around the rocks.*

What do glaciers push in front of them? *Rocks and dirt and anything else that gets in the way*

Do glaciers move quickly or slowly? *They usually move very slowly.*

GLACIER MAZE

Put your collected letters in order here:

GLACIER

Unit 1 Vocabulary Review

1. **Earth science** is the study of our planet.

2. Two things that make Earth special are the **water** that covers most of the surface and the distance we are from the **sun**.

3. Three major events that have changed the way the earth looks are **Creation**, the **Fall** of man, and the Great **Flood**.

4. Noah built a very large boat which we call the **ark**.

5. During the *Ice Age* many parts of the earth were covered with ice.

6. *Glaciers* are large moving sheets of ice that do not completely melt in the summer.

7. *Icebergs* are chunks of ice that fall off of glaciers and float in the water.

8. *Gravity* pulls glaciers down mountains.

Unit 2: Rocks and Minerals

Lesson 8 — Design of the Earth

ANSWER KEY

What are the three main parts of the earth? **Crust, mantle, core**

Which part is the thinnest? **Crust**

Which part is the hottest? Core — specifically **the inner core**

CENTER OF THE EARTH MAZE

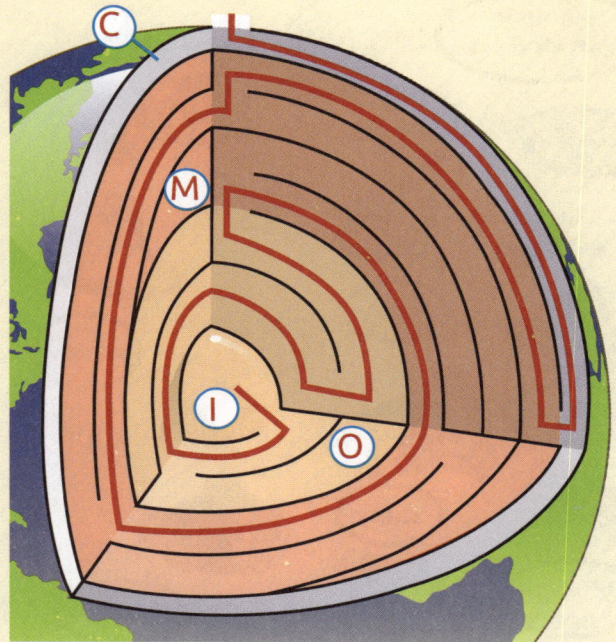

Lesson 9 — Rocks

ANSWER KEY

Where are you likely to find rocks? *If you dig down underneath any place on earth you will find rock.*

How are igneous rocks formed? *Melted rock cools to form igneous rock.*

How are sedimentary rocks formed? *Bits of sand and broken rocks are glued together to form sedimentary rock.*

How are metamorphic rocks formed? *Heat and pressure change igneous or sedimentary rocks into metamorphic rock.*

✏️ CHARACTERISTICS OF ROCKS

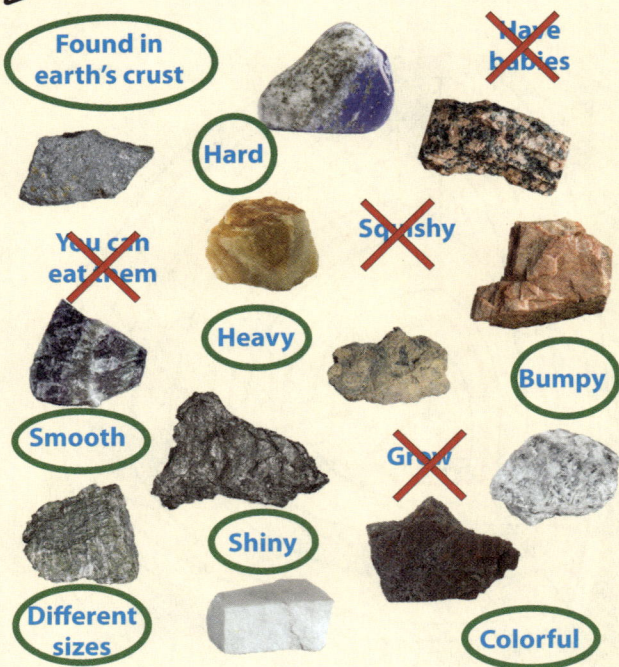

- Found in earth's crust
- ~~Have babies~~
- Hard
- ~~You can eat them~~
- ~~Squashy~~
- Heavy
- Bumpy
- Smooth
- ~~Grow~~
- Shiny
- Different sizes
- Colorful

Lesson 10 — Igneous Rocks

ANSWER KEY

What is lava? *Liquid/melted rock that is outside the earth*

What is one way that lava gets out of the earth's crust? *When a volcano erupts*

What does lava turn into when it cools? *Igneous rock*

✏️ NAME THAT STONE

Obsidian

Lesson 11 — Sedimentary Rocks

ANSWER KEY

What are sedimentary rocks made from? *Bits of sand, rock, and sea shells, glued together*

What event formed most of the sedimentary rocks we see today? *The Great Flood*

✏️ MIXING CONCRETE

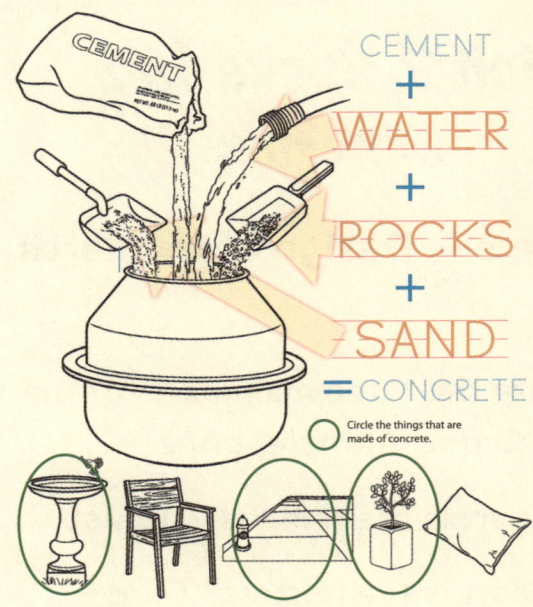

CEMENT
+
WATER
+
ROCKS
+
SAND
= CONCRETE

Circle the things that are made of concrete.

Lesson 12 — Fossils

ANSWER KEY

What is a fossil? *Parts of a dead plant or animal that have been turned to rock*

Do all animals become fossils when they die? *No. The animal must be* covered with mud quickly after it dies. Then the mud must be filled with a liquid that can turn to stone. This does not normally happen to a dead animal.

What kind of rock would you look at if you wanted to find a fossil? *Sedimentary*

🖊 FOSSIL MATCHING

dragonfly

sea turtle

nautilus

criniod

Lesson 13 — Fossil Fuels

ANSWER KEY

Name three types of fossil fuel. *Coal, oil, natural gas*

Why are these things called fossil fuels? *They are found in the same rocks as fossils.*

What is coal made from? *Dead plants*

What is oil made from? *Dead sea creatures*

🖊 MATCH THE FUEL

COAL

GAS

OIL

Lesson 14 — Metamorphic Rocks

ANSWER KEY

What does metamorphic mean? *Changed*

What two things are needed for a rock to change into a metamorphic rock? **Heat and pressure**

Name one kind of metamorphic rock. **Marble**

Have you ever seen a marble statue or floor? **Answers will vary.**

✏️ ROCK REVIEW

1. Rock made from cemented dirt, sand, and rocks is: **Sedimentary**

2. Rock from lava or magma is: **Igneous**

3. Rock that was changed by heat and pressure is: **Metamorphic**

Lesson 15 — Minerals

ANSWER KEY

What are rocks made from? **Minerals**

What are some minerals you found around your house? **Answers will vary.**

Lesson 16 — Identifying Rocks

ANSWER KEY

What are three things you can look at to help you sort rocks? **Color, shininess, crystals**

What does your favorite rock look like? **Answers will vary.**

✏️ SORTING ROCKS

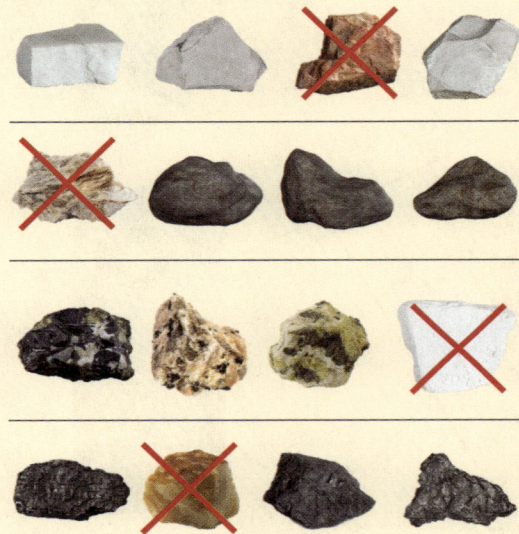

Lesson 17 — Rock Cycle

ANSWER KEY

How can an igneous rock become a metamorphic rock? **Through heat and pressure**

How can a metamorphic rock become a sedimentary rock? **Broken into tiny pieces then glued together with other pieces**

How can a sedimentary rock become an igneous rock? *Melted then cooled*

✏️ COLOR THE ARROWS

Lesson 18 — Gems

ANSWER KEY

What is a gem? *A stone that can be cut to reflect light*

Have you ever seen a diamond? *Answers will vary.*

Why do you think people use gems in jewelry? *They are beautiful and usually valuable.*

✏️ BREASTPLATE COLORING SHEET

How many rows are there? **4**

How many gems are there in each row? **3**

✏️ **Unit 2 Vocabulary Review**

```
S C R Y S T A L O
K C M A G M A V I
M I N E R A L S L
W C O R E Q O O M
E C O A L C W L A
I G N E O U S A N
F O S S I L S V T
I P X G E M S A L
E R O C R U S T E
```

Unit 3: Mountains and Movement

Lesson 19 — The Earth Has Plates

ANSWER KEY

What are large areas of land called? **Continents**

What are large areas of water called? *Oceans*

What continent do you live on? *Answers will vary — most likely North America*

How many continents were there before the flood? **One**

How many continents are there today? **Seven**

✏️ CONTINENTS BEFORE AND AFTER

Before the Flood

Trace the lines to complete the continent of Rodinia, and color the land green.

Write the number of each continent name in the correct space on the map. Then use the code to color each continent.

1–North America	5–Africa	1–Blue	5–Orange
2–South America	6–Australia	2–Red	6–Brown
3–Europe	7–Antarctica	3–Green	7–Purple
4–Asia		4–Yellow	

After the Flood

Lesson 20 — Mountains

ANSWER KEY

What is a mountain? **An area of land that is much taller than the surrounding area**

What is a group of mountains called? **Mountain range**

What is the tallest mountain in the world? **Mount Everest**

✏️ WHAT LIVES IN THE MOUNTAINS

How many trees? **4**

How many animals can you count? **13**

Lesson 21 — Types of Mountains

ANSWER KEY

How do volcanoes form new mountains? **They shoot out lava, ash and rocks that can pile up and form a new mountain.**

What are sand dunes? **Mountains made of sand**

How were some sand dunes formed? **Wind blows sand into one area, causing it to form large hills and even mountains of sand.**

What happened when the earth's plates pushed against each other? **The land in the middle pushed up to form mountains.**

✏️ WHAT TYPE OF MOUNTAIN?

1. **V**   2. **S**   3. **P**

Lesson 22 — Earthquakes

ANSWER KEY

What is an earthquake? **When the ground suddenly shakes**

What is an aftershock? *A smaller earthquake that happens after a big earthquake*

What is a tsunami? *A giant wave caused by an earthquake*

Have you ever felt an earthquake? *Answers will vary.*

✏️ DETECTING EARTHQUAKES

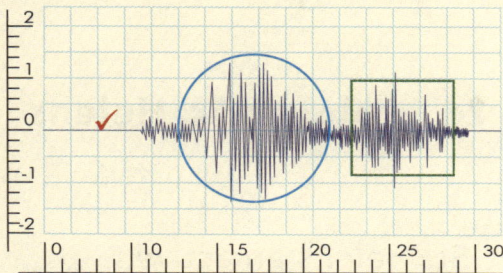

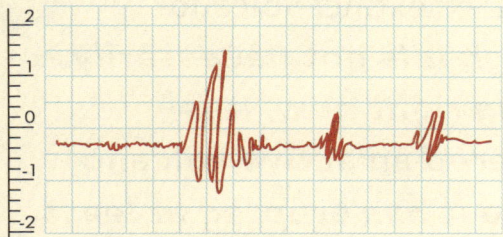

Lesson 23 — Preparing for an Emergency

ANSWER KEY

What are some things that can cause an emergency? *Earthquake, fire, storm*

How can you prepare for an emergency? *Make an emergency kit, practice what to do with your family*

✏️ IN CASE OF EMERGENCY

House on fire

Power is out

Bad storm with lightning

Lesson 24 — Volcanoes

ANSWER KEY

What are some things that come out of volcanoes? *Lava, ash, cinders, steam*

If a volcano has not erupted for a long time but could erupt again, is it dormant or extinct? *Dormant*

✏️ VOLCANO MAZE

Cinders, Ash, Lava, Steam

Lesson 25 — Volcano Types

ANSWER KEY

Why do different volcanoes have different shapes? *The shape is determined by what comes out of the volcano. Mostly lava forms a shield; mostly ash forms a steep short cone, a combination of ash and lava form a taller wider cone.*

Which shape is most common? *Composite — tall, wide cone-shaped*

✏️ VOLCANO MATCH UP

COMPOSITE   CINDER CONE   SHIELD

Lesson 26 — Mount St. Helens

ANSWER KEY

Is Mount St. Helens an active, dormant, or extinct volcano? *Although it is not necessarily erupting right now, it is considered an active volcano because it has erupted at least twice in the past 50 years.*

What happened when it erupted in 1980? *It blew off the top of the mountain. It shot out large amounts of ash, lava, water, and mud. It created large piles of ash and dug out canyons.*

✏️ Unit 3 Vocabulary Review

ANSWER KEY

Continents   Oceans

Rodinia   **E**arthquake

Mountain Ran**g**e   Aftersho**c**k

Sa**n**d Dunes

You should always try to prepare for an **emergency.**

What are three things that often come out of volcanoes? **Ash, Cinders, Steam**

What are three types of volcanoes? **Shield, Cinder Cone, Composite**

If a volcano has erupted recently it is **active**.

If a volcano has not erupted for at least 50 years it is **dormant**.

If a volcano is not expected to ever erupt again it is **extinct**.

Unit 4: Water and Erosion

Lesson 27 — Geysers

ANSWER KEY

Why does water shoot out of a geyser? **Water is heated underground. This causes the water to take up more room and forces the water and steam to shoot out of an opening.**

Where should you go if you want to see Old Faithful? **Yellowstone National Park**

✏ GEYSER TIME

1. (1:00) or 6:00

2. 12:00 or (2:30)

3. 9:00 or (4:00)

4. 3:00 or (5:30)

5. (7:00) or 10:00

6. 11:30 or (8:30)

Lesson 28 — Erosion

ANSWER KEY

What is erosion? **Wearing away of rock by breaking off tiny pieces**

What are the two main things that wear away the rocks? **Water and wind**

Does erosion usually happen quickly or slowly? **Slowly**

What can cause erosion to happen very quickly? **A flood**

✏ **WIND AND WATER**

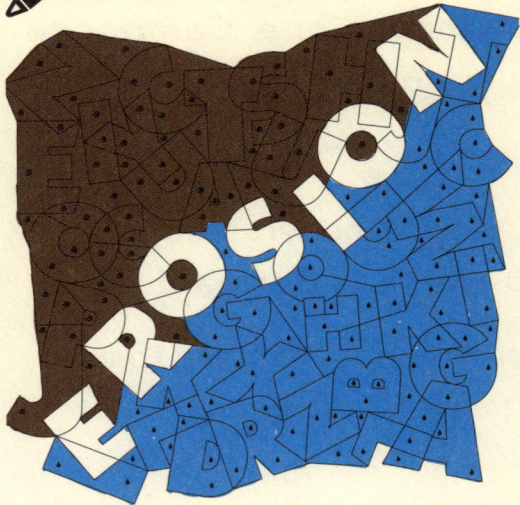

EROSION

Lesson 29 — Landslides

ANSWER KEY

What force pulls everything down? *Gravity*

Is an avalanche more likely to happen in the summer or the winter? *Winter, since an avalanche is moving snow and ice*

✏ **GOING DOWN THE HILL**

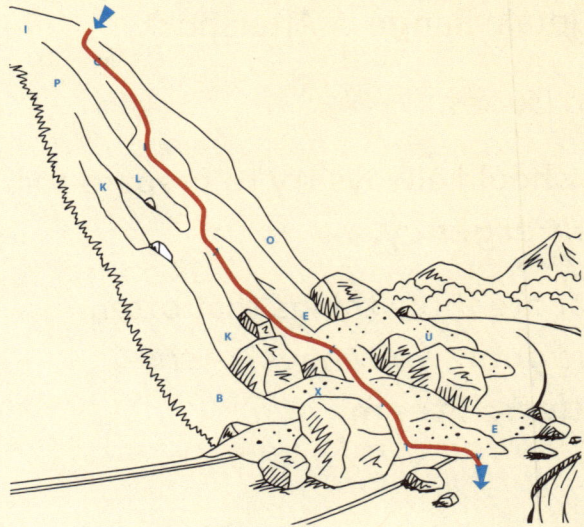

Gravity

Lesson 30 — Stream Erosion

ANSWER KEY

How is stream erosion different from rain or wind erosion? *The water in a stream is moving much faster. Quickly moving water breaks off more bits of rock and dirt so stream erosion is faster.*

Why does water flow downhill? *Gravity pulls it down the hill.*

Does water flow quickly or slowly down a steep hill? *The steeper the hill the faster the water flows.*

Lesson 31 — Soil

ANSWER KEY

What is soil made from? *Bits of sand, clay, and dead plants*

Where do the bits of sand and clay come from? *They are broken off of rocks by erosion*

Why is soil important? *It provides a place for plants to grow and provides food and water for plants.*

✏️ THINGS THAT MAKE SOIL

```
R W O R M S A O O
O R A T R S L W M
C R M W S D W S N
K A W I W M A R I
S R A N P S T O O
L O D D T A E O A
O E O S N I R T I
A N I M A L S S R
P L A N T S L M T
```

Lesson 32 — Grand Canyon

ANSWER KEY

How was the Grand Canyon formed? *Rushing floodwaters eroded away large amounts of rock.*

How do we know that the area was covered with water in the past?

There are many fossils of sea creatures in the sedimentary rock layers of the canyon.

✏️ WHAT MADE THE GRAND CANYON?

Water left over from the *Flood*

Lesson 33 — Caves

ANSWER KEY

What is a cave? *A hollow place underground or in the side of a mountain*

Where might you find stalactites? *On the ceiling of a cave*

Where might you find stalagmites? *On the floor of a cave*

✏️ RHYMES WITH . . .

B ite	*L* ight
WH ite	*R* ight
K ite	*N* ight

Lesson 34 — Rock Collection

✏️ COLLECTING ROCKS

1. How many black rocks did Ronny find? *5*

2. How many brown rocks did he find? **7**

3. Which color of rock did he collect the most of? **Gray**

4. Which color of rock does Ronny have the fewest of? **Green**

5. He found an equal number of which two colors? **Blue and red**

6. How many red and green rocks added together did he find? **3**

7. Challenge: How many rocks did Ronny find altogether? **25**

Lesson 35 — Appreciating Planet Earth

GOD CREATED THE EARTH

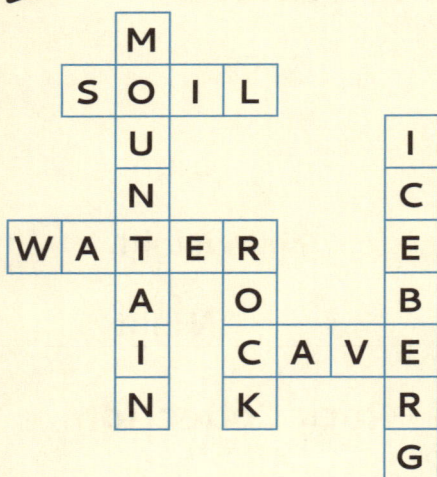

Unit 4 Vocabulary Review

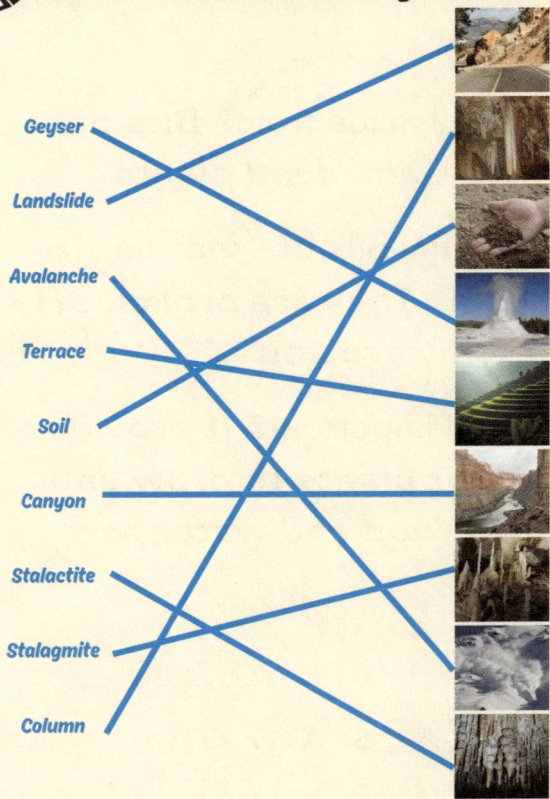

Geyser

Landslide

Avalanche

Terrace

Soil

Canyon

Stalactite

Stalagmite

Column

Photo Credits:

All photos from Getty.com unless specified.

Answers in Genesis: page 42, 43, page 253 (all), page 254 (all), page 257, page 258, page 260, page 262, page 263, page 305, page 307 (top), page 309, page 318, page 321, page 322 (top, middle), page 323 (top), page 324 (all), page 331, page 346, page 349

Author: page 45

European Southern Observatory (ESO): page 196

FEMA: page 25 (Will von Dauster)

istock.com: page 279 (left), page 286, page 334 (bottom)

Jennifer Bauer: page 37, page 187

NASA: page 55, page 80, page 219 (Harrison H. Schmitt), page 176, page 184 (Bill Ingalls), page 195, Page 223 (Bill Ingalls), page 225 (both), page 227, page 228 (Dave Scott), page 230, page 231 (both), page 232, page 233, page 237, page 238,

  Caltech UVA: page 377

  CXC JPL–Caltech STScI NSF NRAO VLA: page 147

  ESA STScI: page 146, page 154, page 221

  Goddard: page 149, page 151, page 167, page 177 (bottom), page 178 (all), page 183, page 189, page 198–199, page 361

Johns Hopkins University Applied Physics Laboratory Carnegie Institution of Washington: page 191

Johns Hopkins University Applied Physics Laboratory Southwest Research Institute: page 215

JPL: page 155 (bottom), page 207 (bottom), page 209 (bottom)

JPL Arizona State University: page 203 (top right, top left)

JPL–Caltech: page 211

JPL–Caltech UCLA: page 152

JPL–Caltech UCLA MPS DLR IDA: page 157 (inset)

JPL–Caltech UMD: page 155 (top)

JPL Hubble Heritage Team STScI AURA: page 131

JPL Space Science Institute: page 191, page 206, page 209 (top)

JPL USGS: page 203 (bottom)

JPL Voyager–ISS Justin Cowart: page 213

NOAA: page 113, page 121 (all), page 122

Pixabay.com: page 279 (right), page 291 (all), page 293 (all but calcium and aluminum), page 295 (second row numbers 2–4) page 307 (bottom), page 319, page 320, page 339 (bottom), page 341, page 343, page 345 (bottom), page 347, page 348 (all), page 353 (all but mountain), page 354 (2, 3, 5, 6, 7, 9)

Thinkstock.com: page 212 (bottom), page 275 (left)

U.S. Geological Survey Photographic Library: page 323 (bottom left)

Vectorstock.com page 315, page 332 (volcano)

Wikimedia:

  CC–BY–2.0: page 45, page 207 (top; Brian Altmeyer)

  CC–BY–SA 2.0 AT: page 160

  CC–BY–3.0: page 124, page 125, page 127, page 164 (top; Asim Patel), page 164 (bottom), page 276, page 325,

  CC–BY–4.0: page 86, page 128, page 177 (top; Reinhold Möller)

  CC BY–SA 3.0 IGO: page 202

  Public Domain: page 161, page 163, page 174,

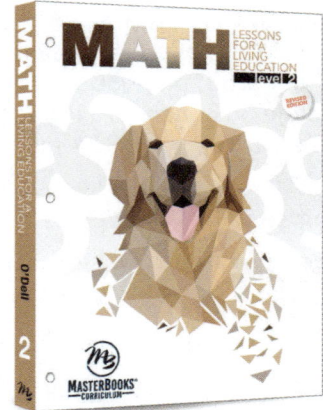

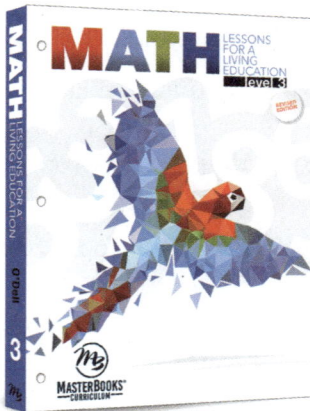

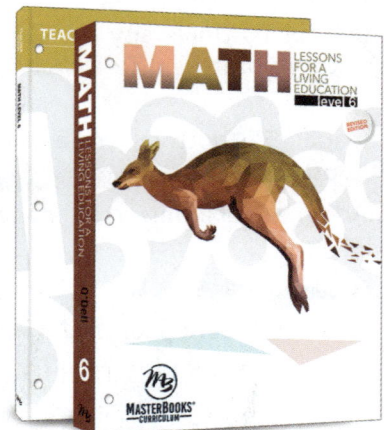